Lecture Notes in Computer Science

T0250772

Commenced Publication in 1973
Founding and Former Series Editors:
Gerhard Goos, Juris Hartmanis, and Jan van Leeuwen

Stefano Leonardi (Ed.)

Algorithms and Models for the Web-Graph

Third International Workshop, WAW 2004
Rome, Italy, October 16, 2004
Proceeedings

 Springer

Volume Editor

Stefano Leonardi
University of Rome "La Sapienza"
Via Salaria 113, 00198 Roma, Italy
E-mail: leon@dis.uniroma1.it

Library of Congress Control Number: 2004113291

CR Subject Classification (1998): F.2, G.2, H.4, H.3, C.2, H.2.8, E.1

ISSN 0302-9743
ISBN 3-540-23427-6 Springer Berlin Heidelberg New York

Springer is a part of Springer Science+Business Media

springeronline.com

© Springer-Verlag Berlin Heidelberg 2004
Printed in Germany

Typesetting: Camera-ready by author, data conversion by Scientific Publishing Services, Chennai, India
Printed on acid-free paper   SPIN: 11333302   06/3142   5 4 3 2 1 0

Preface

This volume contains the 14 contributed papers and the contribution of the distinguished invited speaker Béla Bollobás presented at the 3rd Workshop on Algorithms and Models for the Web-Graph (WAW 2004), held in Rome, Italy, October 16, 2004, in conjunction with the 45th Annual IEEE Symposium on Foundations of Computer Science (FOCS 2004).

The World Wide Web has become part of our everyday life and information retrieval and data mining on the Web is now of enormous practical interest. Some of the algorithms supporting these activities are based substantially on viewing the Web as a graph, induced in various ways by links among pages, links among hosts, or other similar networks.

The aim of the 2004 Workshop on Algorithms and Models for the Web-Graph was to further the understanding of these Web-induced graphs, and stimulate the development of high-performance algorithms and applications that use the graph structure of the Web. The workshop was meant both to foster an exchange of ideas among the diverse set of researchers already involved in this topic, and to act as an introduction for the larger community to the state of the art in this area.

This was the third edition of a very successful workshop on this topic, WAW 2002 was held in Vancouver, Canada, in conjunction with the 43rd Annual IEEE Symposium on Foundations of Computer Science, FOCS 2002, and WAW 2003 was held in Budapest, Hungary, in conjunction with the 12th International World Wide Web Conference, WWW 2003. This was the first edition of the workshop with formal proceedings.

The organizing committee of the workshop consisted of:

Andrei Broder	IBM Research
Guido Caldarelli	INFM, Italy
Ravi Kumar	IBM Research
Stefano Leonardi	University of Rome "La Sapienza"
Prabhakar Raghavan	Verity Inc.

Papers were solicited in all areas of the study of Web graphs, including but not limited to:

- Mathematical models, topology generators, and dynamic properties;
- Algorithms for analyzing Web graphs and for computing graph properties at the Web scale;
- Application of Web graph algorithms to data mining and information retrieval;
- Clustering and visualization;
- Representation and compression;
- Graph-oriented statistical sampling of the Web;
- Empirical exploration techniques and practical systems issues.

The extended abstracts were read by at least three referees each, and evaluated on their quality, originality, and relevance to the symposium. The program committee selected 14 papers out of 31 submissions. The program committee consisted of:

Dimitris Achlioptas	Microsoft Research
Lada Adamic	HP Labs
Jennifer Chayes	Microsoft Research
Fan Chung Graham	UC San Diego
Taher Haveliwala	Stanford University and Google
Elias Koutsoupias	Univ. of Athens
Ronny Lempel	IBM Research
Stefano Leonardi (Chair)	Univ. of Rome "La Sapienza"
Mark Manasse	Microsoft Research
Kevin McCurley	IBM Research
Dragomir Radev	Univ. of Michigan
Sridhar Rajagopalan	IBM Research
Oliver Riordan	Cambridge University
D. Sivakumar	IBM Research
Panayotis Tsaparas	Univ. of Helsinki
Eli Upfal	Brown University
Alessandro Vespignani	Univ. of Paris Sud

WAW 2004, and in particular the invited lecture of Béla Bollobás, was generously supported by IBM. A special thanks is due to Andrei Broder for his effort in disseminating the Call for Papers, to Ravi Kumar for handling the Web site of the – Workshop, and to Debora Donato for her assistance in assembling these proceedings. We hope that this volume offers the reader a representative selection of some of the best current research in this area.

August 2004   Stefano Leonardi
                                                        Program Chair
                                                           WAW 2004

Table of Contents

The Phase Transition and Connectedness in Uniformly Grown Random Graphs

Béla Bollobás[1,2,*] and Oliver Riordan[2,3]

[1] Department of Mathematical Sciences, University of Memphis,
Memphis TN 38152, USA
[2] Trinity College, Cambridge CB2 1TQ, UK
[3] Royal Society Research Fellow, Department of Pure Mathematics and
Mathematical Statistics, University of Cambridge, UK

Abstract. We consider several families of random graphs that grow in time by the addition of vertices and edges in some 'uniform' manner. These families are natural starting points for modelling real-world networks that grow in time. Recently, it has been shown (heuristically and rigorously) that such models undergo an 'infinite-order phase transition': as the density parameter increases above a certain critical value, a 'giant component' emerges, but the speed of this emergence is extremely slow. In this paper we shall present some of these results and investigate the connection between the existence of a giant component and the connectedness of the final infinite graph.

1 Introduction

Recently, there has been a lot of interest in modelling networks in the real world by random graphs. Unlike classical random graphs, many (perhaps most) large networks in the real world evolve in time; in fact they tend to grow in time by the addition of new nodes and new connections. Real-world networks differ from classical random graphs in other important ways (for example, they are often 'scale-free', in the sense of having a power-law degree distribution), and, of course, one cannot expect to model any particular network very accurately, as the real mechanisms involved are not amenable to mathematical analysis. Nevertheless, it is important to model these networks as well as one can, and one general approach is to develop mathematical models for important general features. These models should be simple enough that their properties can be analyzed rigorously. Of course, such models will not be accurate for any given network, but they will give insight into the behaviour of many networks.

One important property of real-world networks is their *robustness*, or resilience to random failures. There are many ways in which one might measure robustness; perhaps the most common is to consider deleting edges or vertices

* Research supported by NSF grant ITR 0225610 and DARPA grant F33615-01-C-1900.

S. Leonardi (Ed.): WAW 2004, LNCS 3243, pp. 1–18, 2004.

from the network at random, and ask whether the network fractures into 'small' pieces, or whether a 'giant component' remains, i.e., a component containing a constant fraction of the initial graph. We shall describe a precise form of this question below.

2 Models

The baseline that any new random graph model should initially be compared with is, and will remain, the classical 'uniform' random graph models of Erdős and Rényi, and Gilbert. Erdős and Rényi founded the theory of random graphs in the late 1950s and early 1960s, setting out to investigate the properties of a 'typical' graph with n vertices and M edges. Their random graph model, $G(n, M)$, introduced in [13], is defined as follows: given $n \geq 2$ and $0 \leq M \leq N = \binom{n}{2}$, let $G(n, M)$ be a graph on n labelled vertices (for example, on the set $[n] = \{1, 2, \ldots, n\}$) with M edges, chosen uniformly at random from all $\binom{N}{M}$ such graphs.

Around the same time that Erdős and Rényi introduced $G(n, M)$, Gilbert [15] introduced a closely related model, $G(n, p)$. Again, $G(n, p)$ is a random graph on n labelled vertices, for example on the set $[n]$. The parameter p is between 0 and 1, and $G(n, p)$ is defined by joining each pair $\{i, j\} \subset [n]$ with an edge with probability p, independently of every other pair. For a wide range of the parameters, for many questions, there is essentially no difference between $G(n, M)$ and $G(n, p)$, where $p = M/N$. Nowadays, $G(n, p)$ is much more studied, as the independence between edges makes it much easier to work with.

Although the definition of $G(n, p)$ has a more probabilistic flavour than that of $G(n, M)$, it was Erdős and Rényi rather than Gilbert who pioneered the use of probabilistic methods to study random graphs, and it is perhaps not surprising that $G(n, p)$ is often known as 'the Erdős-Rényi random graph'. When studying $G(n, p)$, or $G(n, M)$, one is almost always interested in properties that hold for 'all typical' graphs in the model. We say that an event holds *with high probability* or **whp**, if it holds with probability tending to 1 as n, the number of vertices, tends to infinity.

Perhaps the single most important result of Erdős and Rényi about random graphs concerns the emergence of the giant component. Although they stated this result for $G(n, M)$, we shall state it for $G(n, p)$; this is a context in which the models are essentially interchangeable.

For $x > 0$ a constant let

$$t(x) = \frac{1}{x} \sum_{k=1}^{\infty} \frac{k^{k-1}}{k!} \left(x e^{-x} \right)^k. \tag{1}$$

Erdős and Rényi [14] proved the following result.

Theorem 1. *Let $x > 0$ be a constant. If $x < 1$ then* **whp** *every component of $G(n, x/n)$ has order $O(\log n)$. If $x > 1$ then* **whp** *$G(n, x/n)$ has a component with $(1 - t(x) + o(1))n = \Theta(n)$ vertices, and all other components have $O(\log n)$ vertices.*

In other words, there is a 'phase transition' at $x = 1$. This is closely related to the robustness question described vaguely in the introduction, and to the percolation phase transition in random subgraphs of a fixed graph. In this paper, we shall often consider a *random* initial graph, and delete edges (or vertices) independently, retaining each with some probability p, to obtain a random subgraph. The question is, given the (random) initial graph on n vertices, for which values of p does the random subgraph contain a *giant component*, i.e., a component with $\Theta(n)$ vertices? In the context of $G(n, p)$, and in several of the examples we consider, there is no need for this two step construction: if edges of $G(n, p_1)$ are retained independently with probability p_2, the result is exactly $G(n, p_1 p_2)$. In these cases, the robustness question can be rephrased as follows: 'for which values of the edge density parameter is there (**whp**) a giant component?' In the case of $G(n, p)$, the natural normalization is to write $p = x/n$ and keep x fixed as n varies. Thus we see that the classical result of Erdős and Rényi stated above, is exactly a (the first) robustness result of this form.

When the giant component exists, one is often interested in its size, especially near the phase transition. In principle, for $G(n, p)$, the formula (1) above answers this question. More usefully, at $x = 1$ the right-derivative of $t(x)$ is -2, so when $x = 1 + \varepsilon$, the limiting fraction (as $n \to \infty$ with $\varepsilon > 0$ fixed) of vertices in the giant component is $2\varepsilon + o(\varepsilon)$.

2.1   The CHKNS Model

Many growing real-world networks have a number of direct connections that grows roughly linearly with the number of nodes. From now on we shall use graph theoretic terminology, so vertices and edges will correspond to nodes and direct connections between pairs of nodes. (Here we consider undirected connections only.) A very natural model for a growing graph which has on average a constant number of edges per vertex (i.e., in the limit, a constant average degree) was introduced by Callaway, Hopcroft, Kleinberg, Newman and Strogatz [9] in 2001. The model is defined as follows: at each time step, a new vertex is added. Then, with probability δ, two vertices are chosen uniformly at random and joined by an undirected edge. In [9] loops and multiple edges are allowed, although these turn out to be essentially irrelevant. We shall write $G_C^{(n)}(\delta)$ for the n-vertex CHKNS graph constructed with parameter δ. Most of the time, we shall suppress the dependence on n, writing $G_C(\delta)$, to avoid cluttering the notation.

The question considered by Callaway et al in [9] is the following: as the parameter δ is varied, when does $G_C(\delta)$ contain a giant component, i.e., a component containing order n vertices? In other words, what is the equivalent of Theorem 1 for $G_C(\delta)$? In [9], a heuristic argument is given that there is a phase transition at $\delta = \delta_c = 1/8$, i.e., that for $\delta \leq 1/8$ there is no giant component, while for $\delta > 1/8$ there is. Callaway et al also suggest that this phase transition has a particularly interesting form: for $\delta = 1/8 + \varepsilon$, they give numerical evidence (based on integrating an equation, not simply simulating the graph) to suggest that the average fraction of the vertices that lie in the giant component is a function $f_C(\varepsilon)$ which has all derivatives zero at $\varepsilon = 0$. Such a phase transition is called an *infinite-order* transition.

In essence, the question of finding the critical probability for the existence of a giant component in $G_C(\delta)$ had been answered more than a decade before Callaway et al posed the question. As we shall see in section 2.3, a question that turns out to be essentially equivalent was posed by Dubins in 1984, answered partially by Kalikow and Weiss [17] in 1988, and settled by Shepp [20] in 1989.

Dorogovtsev, Mendes and Samukhin [10] analyzed the CHKNS model in a way that, while fairly mathematical, is still far from rigorous. Their methods supported the conjecture of [9], suggesting that indeed the transition is at $\delta_c = 1/8$, and that the phase transition has infinite order.

Before turning to comparison with other models, note that there is a natural slight simplification of the CHKNS model, suggested independently in [11] and [3]. At each stage, instead of adding a single edge between a random pair of vertices with probability δ, for each of the $\binom{n}{2}$ pairs of vertices, add an edge between them with probability $\delta/\binom{n}{2}$, independently of all other pairs. In this way, the number of edges added in one step has essentially a Poisson distribution with mean δ. In the long term, this will make very little difference to the behaviour of the model. The key advantage is that, in the graph generated on n vertices, different edges are now present independently: more precisely, for $\{i,j\} \neq \{i',j'\}$, whether there is an edge (or edges) between i and j is independent of whether there is an edge (or edges) between i' and j'. Note also that the expected number of edges between i and j, $1 \leq i < j \leq n$, is exactly

$$\sum_{k=j}^{n} \delta \binom{k}{2}^{-1} = 2\delta \left(\frac{1}{j-1} - \frac{1}{n} \right). \tag{2}$$

2.2   The Uniformly Grown Random Graph

Although it is not our main focus here, perhaps the most studied growing-graph model is the growth with preferential attachment 'model' of Barabási and Albert, introduced in [1]. The reason for the quotation marks is that the description given by Barabási and Albert is incomplete, and also inconsistent, so their 'model' is not a model in a mathematical sense. Roughly speaking, the Barabási-Albert, or BA, model is defined as follows: an integer parameter m is fixed, and the graph grows by adding one vertex at a time, with each new vertex sending m edges to old vertices, chosen *with probabilities proportional to their degrees*. (This is known as the 'preferential attachment rule'). To prove rigorous results, one must first know exactly what model one is talking about, and this is the reason for the introduction of the *linearized chord diagram* or LCD model in [6]. (See [6] for a description of the problems with the BA model, and [4] for a detailed discussion.) The LCD model is a precisely defined model (one of many) fitting the rough description of Barabási and Albert. It also has an important extra property: although defined as a growing graph process, with rules for how each new vertex attaches to the old graph, it has an equivalent static description, giving the whole n-vertex graph in one go. This makes it much easier to analyze.

The main motivation of Barabási and Albert in [1] was to provide a model explaining the power-law distribution of degrees seen in many real-world networks. They show heuristically that their model does indeed have a power-law degree distribution; this is proved rigorously for the LCD model in [8]. Barabási and Albert note that their new model differs in two fundamental ways from classical uniform models – growth in time, and the preferential attachment rule. They ask whether both these differences are necessary to obtain power-law degree distribution, leading naturally to the study of the following model.

Given an integer m, start with m vertices and no edges. At each time step, add a new vertex to the graph, and join it to a set of m earlier vertices, chosen uniformly at random from among all possible such sets. We shall call this the *growing m-out model*, and write $G_m^{(n)}$, or simply G_m, for the n-vertex graph obtained after $n - m$ steps. Note that this is perhaps the most natural model for a growing graph with (asymptotically) constant average degree.

Barabási and Albert [1] considered G_m briefly, noting that it does not have a power-law degree sequence. Turning to other properties of the BA or LCD models, the question of robustness was considered rigorously in [5] (working with the precisely defined LCD model). It turns out that the LCD model is in some sense 'infinitely robust' in that there is no phase transition: for any $p > 0$, if edges (or vertices) of the $m \geq 2$ LCD graph are retained independently with probability p, there is a giant component, although it may be very small. (Its size is linear in n, but the constant is extremely small when p is small.) Again, it is natural to ask if this striking difference from classical random graphs is due to growth or preferential attachment or both, providing another reason to study the phase transition in growing models, and in particular in G_m. The answer given in [5] is that growth alone is not enough. Much more precise results are given in [7] and [19]; we return to this later.

Just as for the CHKNS model $G_C(\delta)$, there is a natural modification to the growing m-out model G_m that makes it easier to study. Instead of adding exactly m edges from the new vertex, when adding the jth vertex, join it independently to each of the $j - 1$ earlier vertices, joining it to each with probability $m/(j - 1)$. Writing μ instead of m, as there is now no reason for the parameter to be an integer, one obtains a graph on n vertices in which edges are present independently, and for $1 \leq i < j \leq n$ the probability that ij is an edge is

$$\frac{\mu}{j - 1}. \tag{3}$$

We call this graph the *uniformly grown random graph* and denote it by $G_U^{(n)}(\mu)$, or simply $G_U(\mu)$. Note that the model makes perfect sense for $\mu > 1$, but we should write $\min\{\mu/(j - 1), 1\}$ in place of (3). It will turn out that the interesting values of μ are less than 1; also, the results we shall consider do not depend on the presence or absence of the first few edges. Thus we shall not bother with this detail. As we shall see later, and as one can guess by comparing (2) and (3), $G_C(\delta)$ and $G_U(\mu)$ will turn out to be closely related when $\mu \sim 2\delta$.

The modification described above is rather larger than that suggested for the CHKNS model: it destroys the property that the graph is m-out, i.e., that each vertex sends exactly m edges to earlier vertices. As in the CHKNS model edges are always added between random vertices (rather than the new vertex and a random old vertex), there is no corresponding property in the CHKNS model, and the change to independence is a very small change indeed.

2.3 Dubins' Model

In 1984, Dubins proposed a model for an *infinite* inhomogeneous random graph $G_D(\lambda)$ (see [17, 20]). This is the graph on the vertex set $\mathbf{N} = \{1, 2, 3, \dots\}$ in which each edge ij is present independently, and the probability of the edge ij, $1 \le i < j$, is

$$\frac{\lambda}{j}, \tag{4}$$

where $\lambda > 0$ is a real parameter. As before, if $\lambda > 2$ we should write $\min\{\lambda/j, 1\}$ above, but the interest is in smaller values of λ, so we shall not bother. It will come as no surprise that there is a strong connection between $G_D(\lambda)$ and the finite graphs $G_U(\mu)$ and $G_C(\delta)$, when $\lambda \sim \mu \sim 2\delta$. Dubins asked the following question: when $\lambda = 1$, is the graph $G_D(\lambda)$ almost surely *connected*?

At first sight, this question may seem rather strange. For one thing, infinite random graphs are frequently not very interesting. For example, if we take a fixed probability p, $0 < p < 1$, and join each pair $i, j \in \mathbf{N}$ independently with probability p, then with probability 1 we get *the infinite random graph* R. As R is not in fact a random object, and does not depend on p, probabilistic questions about it do not make sense. In fact, R has (in some sense) no structure: for every pair of finite disjoint sets U and W of vertices of R, there are infinitely many other vertices v each of which is joined to every vertex in U and to no vertex in W. Thus, R is trivially connected, having diameter 2. The infinite random graph proposed by Dubins is a very different matter: its structure depends very much on λ, and there are many non-trivial questions one can ask about it, in particular, whether it is connected.

Kalikow and Weiss [17] showed in 1988 that for $\lambda > 1$ the graph $G_D(\lambda)$ is connected (here and from now on, with probability 1 is to be understood), using a very weak form of the classical result of Erdős and Rényi [14] given as Theorem 1 above. They also showed that for $\lambda < 1/4$, $G_D(\lambda)$ is disconnected. They conclude that 'While we are fairly certain that $\frac{1}{4}$ can be replaced by 1, what happens at the critical value $\lambda = 1$ remains for us a mystery.' It is clear that there is some critical value λ_c (a priori, this could be zero or infinity) such that for $\lambda > \lambda_c$, $G_D(\lambda)$ is connected, while for $\lambda < \lambda_c$, $G_D(\lambda)$ is disconnected. Dubins believed that this critical value is equal to 1.

In 1989, Shepp [20] showed that for $\lambda > 1/4$ the graph $G_D(\lambda)$ is connected (with probability 1), establishing that in fact $\lambda_c = 1/4$. He also showed that at $\lambda = 1/4$, $G_D(\lambda)$ is disconnected. A corresponding result for a generalization of the model was proved by Durrett and Kesten [12] in 1990.

3 Connections

It is clear that two of the models considered above are very closely related. In defining the uniformly grown random graph, it makes very little difference if we take j instead of $j - 1$ in (3) (this can actually be formalized, see [3]). After making this change, the only difference between the uniformly grown random graph $G_U(\mu)$, $\mu = \lambda$, and Dubins' graph $G_D(\lambda)$ is that the former is finite and the latter is infinite. In fact, $G_D(\lambda)$ can be viewed as the limit of $G_U(\lambda) = G_U^{(n)}(\lambda)$ as n, the number of vertices, tends to infinity.

In many contexts the $-1/n$ correction term in (2) also makes little difference. Comparing (2) and (3), one might expect $G_C(\delta)$ and $G_U(\mu)$ to behave similarly when μ is (approximately) equal to 2δ. Again, this can be formalized; it was shown by Bollobás, Janson and Riordan [3] that the thresholds for emergence of the giant component in G_C and G_U differ by a factor of *exactly* two, and that the phase transition is infinite-order in one if and only if it is in the other.

The models we consider are related, but what about the questions? For $G_C(\delta)$ or $G_U(\mu)$, we are interested in the question of when (i.e., for which values of the edge-density parameter) there is a giant component. Dubins asked when $G_D(\lambda)$ is connected. Translating to a finite context, the question of whether $G_D(\lambda)$ is connected is the same as the following: is it true that any two vertices i, j will eventually become connected (meaning, joined by a path) if we run the process $G_U(\lambda)$ for long enough? In other words, as n increases, does any fixed pair i, j, of vertices eventually become connected? The giant component question can be rephrased as the following: in the n-vertex graph $G_U(\lambda)$, is the number of connected pairs of order n^2? These two questions are related, but the connection is not obviously a very strong one. Indeed, *a priori*, it is possible that they could have different answers either way round: it might be that any two vertices become connected eventually, but not for a very long time, so the number of connected pairs in the n-vertex graph grows, but very slowly. On the other hand, order n^2 pairs could be connected, but there could be a few pairs that never become connected. It is easy to construct models in which either of these possibilities occurs.

It turns out, however, that in the context of uniformly grown graphs the two questions discussed above *are* very closely related. Indeed, the methods of Kalikow and Weiss [17] and Shepp [20] show that for $\lambda \leq 1/4$, **whp** there is no giant component in $G_U(\lambda)$, while for $\lambda > 1/4$, **whp** there is. The latter result is proved explicitly by Shepp, and then used to immediately deduce connectedness, in the same way that Kalikow and Weiss [17] used the classical Erdős-Rényi giant component result. Thus the question of where the phase transition (defined by the emergence of a giant component) happens in the uniformly grown random graph $G_U(\lambda)$ had already been answered in 1989 – the transition is at $\lambda_c = 1/4$. Recently, Durrett [11] pointed out that the methods of Durrett and Kesten [12] give a corresponding answer for the CHKNS model $G_C(\delta)$. Alternatively, as pointed out in [3], one can compare the models $G_C(\delta)$ and $G_U(\lambda)$, $\lambda \sim 2\delta$, directly: it is easy to show that the critical value δ_c at which phase transition in

$G_C(\delta)$ occurs is given exactly by $\delta_c = \lambda_c/2$. Either way, as Durrett [11] points out, the value $\delta_c = 1/8$ for the point at which the phase transition occurs was (essentially) proved rigorously more than 10 years before Callaway et al asked the question!

4 Slow Emergence

So far we have considered the question of where the phase transition occurs in uniformly grown random graphs, noting that this question was (essentially) answered well before it was asked by Callaway et al. We now turn to the size of the giant component when it exists. Kalikow and Weiss [17] and Shepp [20] did not explicitly address this question, as, for the answer to Dubins question, the size of the giant component is not important. Nevertheless, as we shall note below, a lower bound on the size of the giant component can be read out of the work of Shepp, or the generalization of Durrett and Kesten [12].

Callaway et al [9] heuristically derived a generating function equation for the limiting proportion b_k of vertices in components of size k in the CHKNS graph $G_C(\delta)$, suggesting that the generating function $g_C(x) = \sum_{k\geq 1} x^k b_k$ satisfies the equation

$$g_C'(x) = \frac{1}{2\delta x}\frac{x - g_C(x)}{1 - g_C(x)}. \tag{5}$$

They also presented evidence (from the numerical integration of this equation) that when $\delta = 1/8 + \varepsilon$, the giant component contains asymptotically a fraction $f_C(\varepsilon)$ of the vertices, with

$$f_C(\varepsilon) \sim \exp\left(\alpha\varepsilon^{-\beta}\right),$$

where a best-fit gives $\beta = 0.499 \pm 0.001$.

Dorogovtsev, Mendes and Samukhin [10] analyzed the generating function equation (5) using more mathematical methods that are, nevertheless, not rigorous, presenting interesting arguments suggesting that

$$f_C(\varepsilon) = 2c\exp\left(-\frac{\pi}{2\sqrt{2}}\varepsilon^{-1/2}\right),$$

where $c = 0.295....$ (Their equation C10). Presumably, the assertion of exact equality is not intended, and the suggestion is that this relation holds with $\sim$ instead of =, although this is not clearly stated. It may well be possible to make the methods of [9] and [10] rigorous. Indeed, Durrett made a start in [11], deriving the generating function equation rigorously. He suggested, however, that it would be a thankless task to try to fill in the details in the Dorogovtsev, Mendes, Samukhin proof.

Durrett, using the methods of [12], presented a rigorous proof that the critical value in the CHKNS model is indeed $\delta_c = 1/8$. He also gave a lower bound on the size of the giant component, proving that

$$f_C(\varepsilon) \geq C\varepsilon^2 e^{-3/\varepsilon}. \tag{6}$$

As pointed out independently in [3], Shepp's 1989 proof gives the same type of bound for the uniformly grown graph G_U. Again, one can carry across a bound of this form from one model to the other.

The correct power of ε in the exponent was obtained rigorously in [3], where the following results were proved. We write $C_1(G)$ for the number of vertices in the largest component of a graph G.

Theorem 2. *For any $\eta > 0$ there is an $\varepsilon(\eta) > 0$ such that if $0 < \varepsilon < \varepsilon(\eta)$ then*

$$C_1(G_U^{(n)}(1/4 + \varepsilon)) \geq \exp\left(-\frac{\pi + \eta}{2\sqrt{\varepsilon}}\right) n$$

holds **whp** *as $n \to \infty$.*

Theorem 3. *For any $\eta > 0$ there is an $\varepsilon(\eta) > 0$ such that if $0 < \varepsilon < \varepsilon(\eta)$ then*

$$C_1(G_U^{(n)}(1/4 + \varepsilon)) \leq \exp\left(-\frac{1 - \eta}{2\sqrt{\varepsilon}}\right) n$$

holds **whp** *as $n \to \infty$.*

As noted in [3], and mentioned above, the similarity of the models $G_U(\lambda)$ and $G_C(\lambda/2)$ allows one to deduce that

$$f_C(\varepsilon) = \exp\left(-\Theta(\varepsilon^{-1/2})\right)$$

as $\varepsilon \to 0$, confirming the conjecture of [9] and the heuristics of [10].

5 Path Counting

The proof of Theorem 3 above is based on *path counting*, i.e., counting the expected number of paths in the random graph between a given pair of vertices. This method is often used to provide a lower bound on the critical probability: if the expected number of paths in an n-vertex graph is $o(n^2)$, then on average a proportion $o(1)$ of pairs of vertices can be connected, so the largest component has size $o(n)$ **whp**. Surprisingly, path-counting estimates can be used to obtain upper bounds on the size of the giant component in the models considered here, above the critical probability. This is surprising, because such supercritical graphs have very many paths: the giant component tends to contain exponentially many paths. The key is to form the supercritical graph as a union of a (just) subcritical graph and a graph with few edges, and count paths in the union in the right way.

A path-counting estimate was given by Zhang [21] in 1991, who showed that for vertices $s < t$ in the *critical* infinite graph $G_D(\lambda)$, $\lambda = 1/4$, the probability that s and t are joined by a path is

$$\Theta\left(\frac{\log s}{\sqrt{st}}\right)$$

provided $s \geq (\log t)^{6+\eta}$ for any fixed $\eta > 0$. (He proved this result in the more general situation analyzed by Durrett and Kesten [12]). Durrett [11] gave an outline proof of (a stronger form of) Zhang's result for the finite CHKNS graph G_C (the factor $1/\sqrt{ij}$ was omitted from his Theorem 6; this is clearly just a typographical mistake). Durrett stated that below the critical probability, his method gives an upper bound

$$cs^{-\beta}t^{-1+\beta}, \tag{7}$$

on the probability that s and t are joined by a path, where $\beta = 1/2 - \sqrt{1 - 8\delta}/2$. Here we have translated to the notation of [3].

In [3], a renormalization method was used to prove the result below. Here

$$\beta = \frac{1}{2} - \frac{\sqrt{1 - 4\lambda}}{2}. \tag{8}$$

This corresponds to Durrett's β above, from the connection between $G_C(\delta)$ and $G_D(2\delta)$.

Theorem 4. *For $1 \leq s < t$ and $\lambda < 1/4$ the expected number of s-t paths in $G_D(\lambda)$ is at most*

$$\left(\beta + \beta^2/(1 - 2\beta)\right) s^{-\beta}t^{-1+\beta}. \tag{9}$$

For $s < t$ the expected number of s-t paths in $G_D(1/4)$ is at most

$$\frac{3 + \log s}{4\sqrt{st}}. \tag{10}$$

The second statement is (an explicit) form of Zhang's result, without the restriction on s and t. The first statement is *exactly* Durrett's bound (7) above: Durrett gave the value of c as $\frac{2\delta}{\sqrt{1-8\delta}}$, and one can check that, surprisingly, the constant in front of (9) simplifies to this! (It is not quite clear whether Durrett's bound, whose proof was only outlined in [11], applies to all s and t, or whether there is some mild condition on s and t – the bound is not stated as a separate theorem.) The work in [11] and [3] is independent: Durrett's paper appeared just after the first draft of [3] was completed. The proofs of the path-counting bound are very different; we shall describe the proof in [3] briefly below. Note that while (7) or (9) can be used as the starting point to prove a result like Theorem 3, it is only a starting point. Durrett did not go on to prove such a result.

The basic idea used in [3] to prove (9) is simple; we shall give a sketch here; for full details see [3]. The idea is to write the expected number of paths between vertices s and t as a sum over all possible paths P of the probability that P is present in $G_D(\lambda)$, and then group the paths P in a helpful way. In $G_D(\lambda)$, the probability that an edge ab is present is $\lambda/\max\{a, b\}$, so it turns out that *monotone* paths, i.e., paths in which the vertex numbers increase (or decrease) along the path, are easiest to deal with. Any path P can be written as a zigzag of monotone paths, for example, a monotone increasing path followed by a monotone decreasing path and then another monotone increasing path. The 'zigs' and 'zags' can be grouped together (apart possibly from the first and last; we ignore this complication here): any path P can be obtained from its sequence S of local minima by inserting a single increasing-decreasing zigzag between each successive pair of minima. Grouping paths P by the sequence S, the sum over all paths P corresponding to a fixed S can be bounded by a simple formula. In fact, viewing S (a sequence of vertices) as a path, the sum is (essentially) just the probability that S is present as a path in a different random graph, described below. The sequence $S(P)$ of minima will be shorter than P, except in some trivial cases, so one can apply this reduction repeatedly to bound the expected number of paths.

More concretely, for $\alpha > 0$ and $0 \leq \beta < 1/2$, let $G(\alpha, \beta)$ be the graph on $\{1, 2, \ldots\}$ in which each edge is present independently, and for $1 \leq i < j$ the probability that ij is present is $\alpha i^{-\beta} j^{-1+\beta}$. Note that $G_D(\lambda) = G(\lambda, 0)$. Let $E_{12}(\alpha, \beta)$ be the expected number of paths from vertex 1 to vertex 2 in $G(\alpha, \beta)$. (For simplicity, we consider only paths between vertices 1 and 2: in this case the first and last vertices of the path are always local minima.) Although this is not exactly the presentation in [3], one can easily check, grouping paths P by the sequence $S(P)$ as above, that

$$E_{12}(\alpha, \beta) \leq E_{12}(\alpha', \beta'),$$

where

$$\alpha' = \frac{\alpha^2}{1 - 2(\alpha + \beta)} \text{ and } \beta' = \alpha + \beta.$$

At first sight, it is not clear that this is helpful: we have reduced our problem to an equally difficult problem. However, this transformation does take us much closer to the solution. Indeed, let $(\alpha_0, \beta_0) = (\lambda, 0)$, and for $k \geq 0$ let

$$(\alpha_{k+1}, \beta_{k+1}) = \left(\frac{\alpha_k^2}{1 - 2(\alpha_k + \beta_k)}, \alpha_k + \beta_k \right). \tag{11}$$

Then, as shown in [3], if $\lambda \leq 1/4$ the α_k tend to zero, and the β_k increase to a limit $\beta \leq 1/2$ given by (8). Using the fact that every path is eventually counted after applying the operation $S(P)$ enough times, one can deduce Theorem 4. (This is only an outline; as mentioned earlier, the details are in [3].)

More recently, a much more precise result has been obtained by a different method, namely direct comparison with a 'continuous-type branching process'. Riordan [19] proved the result below concerning $G_D^{(n)}(\lambda)$, the n-vertex version of the Dubins graph. This is the same as $G_U^{(n)}$, except that j is taken instead of $j - 1$ in (3). As noted in [3], this does not affect the result. We shall write the result for G_U, and adapt the notation for consistency with that above.

Theorem 5. *If $\lambda \leq 1/4$ is constant, then $C_1(G_U^{(n)}(\lambda)) = o(n)$ holds* **whp** *as $n \to \infty$. There is a function f_U such that if $\lambda = 1/4 + \varepsilon$, where ε is a positive* constant, then $C_1(G_U^{(n)}(\lambda)) = (f_U(\varepsilon) + o(1))n$ holds **whp** as $n \to \infty$. Furthermore,

$$f_U(\varepsilon) = \exp\left(-\frac{\pi}{2}\frac{1}{\sqrt{\varepsilon}} + O(\log(1/\varepsilon))\right) \tag{12}$$

as $\varepsilon \to 0$ from above.

Once again, as noted in [19], the result carries over to the CHKNS model, proving rigorously that

$$f_C(\varepsilon) = \exp\left(-\frac{\pi}{2\sqrt{2}}\frac{1}{\sqrt{\varepsilon}} + O(\log(1/\varepsilon))\right).$$

Although these results seem (and are!) very precise, they are rather far from giving us the asymptotic values of the functions $f_U(\varepsilon)$ and $f_C(\varepsilon)$.

6   Other Models

So far we have presented detailed results for the models G_D, G_U and G_C, which are all very closely related. We mentioned earlier that the difference between the growing m-out model G_m and the uniformly grown random graph G_U is a significant one, even though individual edge probabilities are the same in the two models. This is shown by the fact that the phase transition occurs at a different point in the two models. For the model G_m, we cannot vary the parameter m continuously, so we return to the formulation where the random graph is generated, and then edges are deleted, each edge being retained with probability p, independently of the other edges. In $G_U(\mu)$, such an operation has the same effect as multiplying the parameter μ by p. Thus, if the dependence in the m-out model were not important, we would expect a transition at $pm = 1/4$, i.e., at $p = 1/(4m)$. It was shown in [7] that this is *not* what happens: there is a phase transition, indeed an infinite-order one, but it occurs at a different critical probability. The value of this critical probability was found exactly in [7], and bounds analogous to Theorems 2 and 3 were proved. Recently, Riordan [19] proved considerably stronger results, superceding those in [7]: we state these results here, with slightly changed notation.

Theorem 6. *Let* $m \geq 2$ *and* $0 < p < 1$ *be fixed, and set*

$$p_c = p_c(m) = \frac{1}{2}\left(1 - \sqrt{\frac{m-1}{m}}\right) = \frac{1}{4m} + \frac{1}{16m^2} + O(m^{-3}). \tag{13}$$

Let $G_m^{(n)}(p)$ *be the random subgraph of the uniformly grown m-out random graph* $G_m^{(n)}$ *obtained by retaining each edge independently with probability p. Firstly, if $p \leq p_c$, then $C_1(G_m^{(n)}(p)) = o(n)$ holds **whp** as $n \to \infty$. Secondly, there exists a function $f_m(\varepsilon) > 0$ such that if $\varepsilon > 0$ is constant, then $C_1(G_m^{(n)}(p_c+\varepsilon)) = (f_m(\varepsilon) + o(1))n$ holds **whp** as $n \to \infty$. Furthermore,*

$$f_m(\varepsilon) = \exp\left(-\frac{\pi}{2(m(m-1))^{1/4}}\frac{1}{\sqrt{\varepsilon}} + O(\log(1/\varepsilon))\right) \qquad (14)$$

as $\varepsilon \to 0$ from above with m fixed.

We mentioned earlier that for the LCD model, a precisely defined member of the Barabási-Albert class of models, the critical probability is 0: for the $m \geq 2$ LCD model, whenever edges are retained independently with probability $p > 0$, there is a giant component. As $n \to \infty$ with p fixed, the limiting proportion of vertices in the giant component is given by some positive function $f_L(m, p)$. Although this component is 'giant', it is also very small: it was shown in [5] that for m fixed, $f_L(m, p)$ is at most $\exp(-\Theta(1/p))$, and at least $\exp(-\Theta(1/p^2))$. It turns out that the upper bound is correct, as shown by Riordan [19]. The reader is referred to [19] or [6] for the definition of the model, which is not our main focus here.

Theorem 7. *Let $m \geq 2$ and $0 < p < 1$ be fixed, and consider the subgraph G of the n-vertex, parameter m LCD graph obtained by retaining edges independently with probability p. There is a constant $f_L(m, p) > 0$ such that* **whp** *as $n \to \infty$ the largest component of G has $(f_L(m, p) + o(1))n$ vertices. Furthermore, as $p \to 0$ with m fixed we have*

$$\Omega\left(p^2\left(\frac{m-1}{m+1}\right)^{\frac{1}{2p}}\right) = f_L(m, p) = O\left(\left(\frac{m-1}{m+1}\right)^{\frac{1}{2p}}\right). \qquad (15)$$

In particular, as $p \to 0$ with m fixed,

$$f_L(m, p) = \exp\left(-\frac{c_m}{p} + O(\log(1/p))\right) = \exp\left(-\frac{c_m + o(1)}{p}\right), \qquad (16)$$

where

$$c_m = \frac{1}{2}\log\left(\frac{m+1}{m-1}\right) = \frac{1}{m} + \frac{1}{3m^3} + O(m^{-5}). \qquad (17)$$

So far, we have considered the random subgraph of a given (random) graph obtained by deleting edges randomly. In many examples, it is nodes that may fail rather than connections between them. Theorems 6 and 7 hold *mutatis mutandis* if vertices are deleted independently with probability $1 - p$ rather than edges – the only difference is an additional factor of p in the bounds on the size of the giant component.

7   The Window of the Phase Transition

So far, we have considered the size $C_1(G)$ of the largest component in various n-vertex random graphs, asking for the limit of $C_1(G)/n$ as $n \to \infty$ with the relevant edge-density parameter fixed. When this limit is zero, this question is rather crude. We are effectively classifying (nice, i.e., not oscillating) functions of n as either $\Theta(n)$ or $o(n)$; in the latter case it is natural to ask for more detailed results. Going back to $G(n,p)$, Erdős and Rényi [14] showed that for $p = x/n$, $x < 1$, we have $C_1(G(n, x/n)) = O(\log n)$ **whp**. Also, when $x = 1$, they showed that $C_1(G(n, 1/n)) = \Theta(n^{2/3})$. (There is an error in their paper here; they claimed the second result for $p = x/n$, $x \to 1$. Actually, throughout they stated their results for $G(n, M)$ – we have translated to $G(n,p)$ as for this type of result the two models are equivalent.) Of course, there is no reason to set p to be exactly x/n, x constant. To see in detail what is happening at the phase transition, one should look 'inside the window', choosing x to depend on n in the right way.

Bollobás [2] was the first to show, in 1984, that for $G(n,p)$, $p = (1+\varepsilon(n))/n$, in the window $|\varepsilon|$ is about $n^{-1/3}$; in 1990, Łuczak [18] proved more precise results. In 1994, Janson, Knuth, Łuczak and Pittel [16] used generating functions to prove very detailed results about phenomena inside the window. In summary, writing $p = (1 + a(n)n^{-1/3})/n$, if $a(n)$ is bounded as $n \to \infty$, then there are several similarly large components, each having size $O(n^{2/3})$. If $a(n) \to \infty$, then 'the' giant component has already emerged: **whp** there is one component which is much larger than the others (the ratio of the sizes tends to infinity). Also, if one increases the value of p by adding edges to the existing graph appropriately, this component remains the largest component from then on.

Thus, in $G(n,p)$, the giant component becomes clearly identifiable when it has $\omega(n^{2/3})$ vertices. Here, as usual, the notation $f = \omega(g)$ means that $f/g \to \infty$, the limit, as all limits here, being as $n \to \infty$. What is the corresponding result for the uniformly grown graphs considered here? We believe that the answer is about $n^{1/2}$ and, as we shall describe shortly, that this is closely related to the question of when the infinite graph becomes connected.

On the one hand, as noted in [3], setting $\lambda = 1/4$, the first few components of $G_U^{(n)}(\lambda)$ are likely to have similar sizes (within constant factors), and these sizes are presumably $n^{1/2+o(1)}$. (An upper bound $O(n^{1/2}/\log n)$ is given by Durrett [11] for the expected size of the component containing vertex 1, and this seems likely to be fairly close to the truth.)

For the giant component to have size between order $n^{1/2}$ and order n, in the finite graph $G_U^{(n)}(\lambda)$ we must allow λ to depend on n, setting $\lambda = (1 + \varepsilon(n))/4$. This corresponds to taking $p = (1 + \varepsilon(n))/n$ in $G(n,p)$. Unfortunately, this destroys the 'nesting' property, that we may construct the graphs $G_U^{(n)}(\lambda)$ (in the λ/j rather than $\lambda/(j-1)$ variant) simultaneously for all n, as subgraphs induced by the first n vertices of $G_D(\lambda)$. To preserve this property, for $1 \le i < j$ the probability that the edge ij is present should be taken to be

$$\frac{1 + \varepsilon(j)}{4j},$$

where $\varepsilon(j)$ is some function tending to zero. We write G_ε for the infinite graph obtained in this way, and $G_\varepsilon^{(n)}$ for the subgraph induced by its first n vertices. If the function ε is well behaved, the largest component in $G_U^{(n)}(1/4 + \varepsilon(n))$ should be similar in size to the largest component in $G_\varepsilon^{(n)}$. From now on we work with the latter formulation.

It is easy to give a simple heuristic argument suggesting that if the (suitably nice) function ε is chosen so that the largest component of $G_\varepsilon^{(n)}$ has order $n^{1/2}$ or larger, then the infinite graph G_ε will be connected. Indeed, consider the component $C^t = C^t(n)$ of $G_\varepsilon^{(n)}$ containing the vertex t. This component is born at time t and, as each new vertex is added, it may or may not grow. The component C^t increases in size as we pass from n to $n+1$ if and only if the new vertex added is joined to some vertex in $C^t(n)$. When it does increase, the number of vertices added is 1 (the new vertex) plus the number of vertices in other components of $G_\varepsilon^{(n)}$ to which the new vertex is joined. As long as $C^t(n)$ is not very large, the probability that the new vertex joins to it is essentially proportional to $|C^t(n)|$. If we exclude the case where the new vertex joins to the giant component, the expected number of vertices in other components that the new vertex joins to is roughly constant. This suggests that there is a constant a, depending on ε, such that, until $C^t(n)$ joins the largest component, the expected increase in $|C^t(n)|$ is $(a + o(1))|C^t(n)|$. Taking a 'mean-field' approximation, we expect $C^t(n)$ to grow roughly as $(n/t)^a$, until this component joins the giant component.

The heuristic argument just given applies to $t = 1$, suggesting that the largest component (which may not be that containing vertex 1, but will presumably contain some early vertex) will have size roughly n^a, so $a \geq 1/2$ by assumption. Now, as each vertex is added, the probability that it sends edges to both $C^t(n)$ and $C^1(n)$ is roughly n^{2a-2}/t^a. As $2a - 2 \geq -1$, the sum of these probabilities diverges so, with probability 1, eventually the vertices 1 and t should be connected by a path.

The argument above is heuristic, and may well break down when $a = 1/2$. Nevertheless, it seems likely that if $\varepsilon(n)$ is chosen as a 'nice' function so that the size of $C^1(n)$ (or $C_1(G_\varepsilon^{(n)})$) grows at least as fast as $n^{1/2}$, then G_ε will be connected with probability 1. Incidentally, this suggests that the connectedness of G_D and the existence of a giant (meaning order n) component in G_U are not quite as closely related as we have so far suggested – order n is not the relevant order.

Turning to the emergence of the giant component, the same heuristic suggests that, if a above is strictly greater than $1/2$, then the giant component will be much larger than all other components. Indeed, if vertex t is not yet in the giant component, then we expect $C^t(n)$ to contain roughly $(n/t)^a$ vertices, and when the next vertex is added, C^t joins the giant component with probability around n^{2a-2}/t^a. The sum of these probabilities becomes large when n exceeds $O(t^{a/(2a-1)})$. Turning this round, in the n-vertex graph we expect only components born after time around $t_0 = n^{(2a-1)/a}$ to have a significant chance of *not* having joined the giant component. Such components have size around $(n/t_0)^a = n^{1-a}$, which is much smaller than n^a. This suggests that in $G_\varepsilon^{(n)}$, the giant component will be readily identifiable (i.e., exceed the second largest in size by more than a constant factor) when its size significantly exceeds $n^{1/2}$.

8 Conjectures and Open Problems

Let us conclude with two conjectures and three problems. The conjectures and first problem are essentially a challenge to make the heuristics given above precise. We consider $G_\varepsilon^{(n)}$ as defined in the previous section, where the function $\varepsilon(j)$ should be 'nice', i.e., tending to zero 'smoothly', rather than oscillating in some way.

Conjecture 1. Let $\gamma > 0$ be an arbitrary positive constant. If a 'nice' function $\varepsilon(j)$ is chosen so that **whp** the largest component of $G_\varepsilon^{(n)}$ contains at least $n^{1/2+\gamma}$ vertices, then **whp** the second largest component of $G_\varepsilon^{(n)}$ contains at most $n^{1/2}$ vertices.

Our second conjecture will be the converse, stating roughly that if the largest component has size $n^{1/2-\gamma}$, then the largest few components all have similar sizes. In this case, as noted in [3], we believe that the relevant choice of ε is a negative constant, taking us back to G_D, and we can make a more specific conjecture. Here $\beta(\lambda)$ is given by (8).

Conjecture 2. Let $k \geq 1$ and $0 < \lambda < 1/4$ be constant, and let $\mathbf{X} = (X_1, X_2, \ldots, X_k)$ be the sizes of the k largest components of the subgraph of $G_D(\lambda)$ induced by the first n vertices, listed in decreasing order of size. Then $n^{-\beta(\lambda)}\mathbf{X}$ converges in distribution to $\mathbf{Y}$ as $n \to \infty$, where $\mathbf{Y}$ has a non-degenerate distribution on $\{(y_1, y_2, \ldots, y_k) \in \mathbf{R}^k : y_1 > y_2 > \cdots > y_k > 0\}$.

This conjecture was made as a remark in [3].

Turning to the infinite graph G_ε, it seems likely that for nice functions, this is connected when $\varepsilon(j)$ is large enough to ensure a component in $G_\varepsilon^{(n)}$ of order around $n^{1/2}$.

Problem 1. For which (nice) functions $\varepsilon(j)$ is the infinite graph G_ε connected with probability 1? Are these the same functions for which the largest component of $G_\varepsilon^{(n)}$ has size at least $f(n)$, for some function $f(n) = O^*(n^{1/2})$?

Here the O^* notation hides factors bounded by $(\log n)^C$ for some constant; the statement above is rather informal, as we do not wish to make a precise conjecture. The strongest possible conjecture would be that for nice functions $\varepsilon(j)$, the infinite graph is connected if and only if $g(n)$, the expected size of the largest component in $G_\varepsilon^{(n)}$, is such that $\sum_n g(n)^2/n^2$ is divergent. This is what our heuristic suggests, but such a strong statement seems unlikely to be true. The weak statement that when $g(n) \leq n^{1/2-\gamma}$ the infinite graph is disconnected, and when $g(n) \geq n^{1/2+\gamma}$ it is connected, seems very likely to be true.

Finally, we finish with two rather different problems, the first of which may well be very hard.

Problem 2. Prove an asymptotic formula for the size of the giant component just above the phase transition in $G_C(\delta)$ or $G_U(\delta)$. In other words, determine $f_U(\varepsilon)$ and/or $f_C(\varepsilon)$ to within a factor tending to 1 as $\varepsilon \to 0$.

Knowing of an example, $G(n,p)$, where the giant component first becomes identifiable at around size $n^{2/3}$, and believing that G_U provides an example where this happens at around size $n^{1/2}$, the following question is natural.

Problem 3. Are there natural random graph models in which the giant component first becomes readily identifiable when it contains roughly n^α vertices for values of α other than 2/3 and 1/2, in particular, for α very close to 0 or 1?

References

1. A.-L. Barabási and R. Albert, Emergence of scaling in random networks, *Science* **286** (1999), 509–512.
2. B. Bollobás, The evolution of random graphs. *Trans. Amer. Math. Soc.* **286** (1984), 257–274.
3. B. Bollobás, S. Janson and O. Riordan, The phase transition in the uniformly grown random graph has infinite order, to appear in *Random Structures and Algorithms*.
4. B. Bollobás and O. Riordan, Mathematical results on scale-free random graphs, in Handbook of Graphs and Networks, Stefan Bornholdt and Heinz Georg Schuster (eds.), Wiley-VCH, Weinheim (2002), 1–34.
5. B. Bollobás and O. Riordan, Robustness and vulnerability of scale-free random graphs, *Internet Mathematics* **1** (2003), 1–35.
6. B. Bollobás and O. Riordan, The diameter of a scale-free random graph, *Combinatorica* **24** (2004), 5–34.
7. B. Bollobás and O. Riordan, Slow emergence of the giant component in the growing m-out graph, submitted. (Preprint available from http://www.dpmms.cam.ac.uk/∼omr10/.)
8. B. Bollobás, O. Riordan, J. Spencer and G. Tusnády, The degree sequence of a scale-free random graph process, *Random Structures and Algorithms* **18** (2001), 279–290.
9. D.S. Callaway, J.E. Hopcroft, J.M. Kleinberg, M.E.J. Newman and S.H. Strogatz, Are randomly grown graphs really random?, *Phys. Rev. E* **64** (2001), 041902.
10. S.N. Dorogovtsev, J.F.F. Mendes and A.N. Samukhin, Anomalous percolation properties of growing networks, *Phys. Rev. E* **64** (2001), 066110.
11. R. Durrett, Rigorous result for the CHKNS random graph model, Proceedings, Discrete Random Walks 2003, Cyril Banderier and Christian Krattenthaler, Eds. *Discrete Mathematics and Theoretical Computer Science* **AC** (2003), 95–104. http://dmtcs.loria.fr/proceedings/
12. R. Durrett and H. Kesten, The critical parameter for connectedness of some random graphs, in *A Tribute to Paul Erdős* (A. Baker, B. Bollobás and A. Hajnal, eds), Cambridge Univ. Press, Cambridge, 1990, pp 161–176,
13. P. Erdős, and A. Rényi, On random graphs I., *Publicationes Mathematicae Debrecen* **5** (1959), 290–297.
14. P. Erdős and A. Rényi, On the evolution of random graphs, *Magyar Tud. Akad. Mat. Kutató Int. Közl.* **5** (1960), 17–61.
15. E.N. Gilbert, Random graphs, *Annals of Mathematical Statistics* **30** (1959), 1141–1144.
16. S. Janson, D.E. Knuth, T. Łuczak and B. Pittel, The birth of the giant component, *Random Structures and Algorithms* **3** (1993), 233–358.

17. S. Kalikow and B. Weiss, When are random graphs connected?, *Israel J. Math.* **62** (1988), 257–268.
18. T. Łuczak, Component behavior near the critical point of the random graph process, *Random Structures and Algorithms* **1** (1990), 287–310.
19. O. Riordan, The small giant component in scale-free random graphs, to appear in Combinatorics, Probability and Computing.
20. L.A. Shepp, Connectedness of certain random graphs. *Israel J. Math.* **67** (1989), 23–33.
21. Y. Zhang, A power law for connectedness of some random graphs at the critical point, *Random Structures and Algorithms* **2** (1991), 101–119.

Analyzing the Small World Phenomenon Using a Hybrid Model with Local Network Flow (Extended Abstract)

Reid Andersen, Fan Chung, and Lincoln Lu*

University of California, San Diego

Abstract. Randomly generated graphs with power law degree distribution are typically used to model large real-world networks. These graphs have small average distance. However, the small world phenomenon includes both small average distance and the clustering effect, which is not possessed by random graphs. Here we use a hybrid model which combines a global graph (a random power law graph) with a local graph (a graph with high local connectivity defined by network flow). We present an efficient algorithm which extracts a local graph from a given realistic network. We show that the hybrid model is robust in the sense that for any graph generated by the hybrid model, the extraction algorithm approximately recovers the local graph.

1 Introduction

The small world phenomenon usually refers to two distinct properties— *small average distance* and the *clustering effect*— that are ubiquitous in realistic networks. An experiment by Stanley Milgram [11] titled "The small world problem" indicated that any two strangers are linked by a short chain of acquaintances. The clustering effect implies that any two nodes sharing a neighbor are more likely to be adjacent.

To model networks with the small world phenomenon, one approach is to utilize randomly generated graphs with power law degree distribution. This is based on the observations by several research groups that numerous networks, including Internet graphs, call graphs and social networks, have a *power law* degree distribution, where the fraction of nodes with degree k is proportional to $k^{-\beta}$ for some positive exponent β (See [12] for an extensive bibliography). A random power law graph has small average distances and small diameter. It was shown in [3] that a random power law graph with exponent β, where $2 < \beta < 3$, almost surely has average distance of order $\log \log n$ and has diameter of order $\log n$.

In contrast, the clustering effect in realistic networks is often determined by local connectivity and is not amenable to modeling using random graphs. A previous approach to modelling the small world phenomenon was to add random edges to an underlying graph like a grid graph (see Watts and Strogatz

---

* Research supported in part by NSF Grants DMS 0100472 and ITR 0205061.

[14]). Kleinberg [10] introduced a model where an underlying grid graph G was augmented by random edges placed between each node u, v with probability proportional to $[d_G(u,v)]^{-r}$ for some constant r. In Kleinberg's model and the model of Watts and Strogatz, the subgraphs formed by the random edges have the same expected degree at every node and do not have a power law degree distribution. Fabrikant, Koutsoupias and Paradimitriou [6] proposed a model where vertices are coordinates in the Euclidean plane and edges are added by optimizing the trade-off between Euclidean distances and "centrality" in the network. Such grid-based models are quite restrictive and far from satisfactory for modeling webgraphs or biological networks.

In [4] Chung and Lu proposed a general hybrid graph model which consists of a global graph (a random power law graph) and a highly connected local graph. The local graph has the property that the endpoints of every edge are joined by at least l edge-disjoint paths each of length at most k, for some fixed parameters k and l. It was shown that these hybrid graphs have average distance and diameter of order $O(\log n)$ where n is the number of vertices.

In this paper, we consider a new notion of local connectivity that is based on network flow. Unlike the problem of finding short disjoint paths, the local flow connectivity can be easily computed using techniques for the general class of fractional packing problems. The goal is to partition a given real-world network into a global subgraph consisting of "long edges" providing small distances and a local graph consisting of "short edges" providing local connections. In this paper, we give an efficient algorithm which extracts a highly connected local graph from any given real world network. We demonstrate that such recovery is robust if the real world graph can be approximated by a random hybrid graph. Namely, we prove that if G is generated by the hybrid graph model, our partition algorithm will recover the original local graph with a small error bound.

2   Preliminaries

2.1   Random Graphs with Given Expected Degrees

We consider a class of random graphs with given expected degree sequence $\mathbf{w} = (w_1, w_2, \ldots, w_n)$. The probability that there is an edge between any two vertices v_i and v_j is $p_{ij} = w_i w_j \rho$, where $\rho = (\sum w_i)^{-1}$. We assume that $\max_i w_i^2 < \sum_k w_k$ so that $p_{ij} \leq 1$ for all i and j. It is easy to check that vertex v_i has expected degree w_i. We remark that the assumption $\max_i w_i^2 < \sum_k w_k$ implies that the sequence w_i is graphical [5], except that we do not require the $\{w_i\}$ to be integers. We note that this model allows a non-zero probability for self-loops. The expected number of loops is quite small (of lower order) in comparison with the total number of edges.

We denote a random graph with expected degree sequence $\mathbf{w}$ by $G(\mathbf{w})$. For example, the typical random graph $G(n,p)$ on n vertices with edge probability p is a random graph with expected degree sequence $\mathbf{w} = (pn, pn, \ldots, pn)$. For a subset S of vertices, we define $\mathrm{Vol}(S) = \sum_{v_i \in S} w_i$, and $\mathrm{Vol}_k(S) = \sum_{v_i \in S} w_i^k$. We let d denote the average degree $Vol(G)/n$, and let $\tilde{d}$ denote the second order average degree $\mathrm{Vol}_2(G)/\mathrm{Vol}(G)$. We also let m denote the maximum weight among the w_i. The main results of this paper apply to $G(\mathbf{w})$ and are stated in terms of d, $\tilde{d}$, and m. We will mostly be interested in the special case where $G(\mathbf{w})$ is a random power law graph.

2.2   Random Power Law Graphs

A random power law graph $M(n, \beta, d, m)$ is a random graph $G(\mathbf{w})$ whose expected degree sequence $\mathbf{w}$ is determined by the following four parameters.

- n is the number of vertices.
- $\beta > 2$ is the power law exponent.
- d is the expected average degree.
- m is the maximum expected degree and $m^2 = o(nd)$.

We let the i-th vertex v_i have expected degree

$$w_i = ci^{-\frac{1}{\beta-1}},$$

for $i_0 \leq i \leq i_0 + n$, for some c and i_0 (to be chosen later). It is easy to compute that the number of vertices of expected degree between k and $k + 1$ is of order $c'k^{-\beta}$ where $c' = c^{\beta-1}(\beta - 1)$, as required by the power law. To determine c, we consider

$$\mathrm{Vol}(G) = \sum_i w_i = \sum_{i \geq i_0} ci^{\frac{1}{\beta-1}}$$

$$\approx c\frac{\beta - 1}{\beta - 2}n^{1-\frac{1}{\beta-1}}.$$

Here we assume $\beta > 2$. Since $nd \approx \mathrm{Vol}(G)$, we choose

$$c = \frac{\beta - 2}{\beta - 1}dn^{\frac{1}{\beta-1}}, \tag{1}$$

$$i_0 = n(\frac{d(\beta - 2)}{m(\beta - 1)})^{\beta-1}. \tag{2}$$

Values of $\tilde{d}$ for random power law graphs are given below (see [3]).

$$\tilde{d} = \begin{cases} (1 + o(1))d\frac{(\beta-2)^2}{(\beta-1)(\beta-3)} & \text{if } \beta > 3. \\ (1 + o(1))\frac{1}{2}d\ln\frac{2m}{d} & \text{if } \beta = 3. \\ (1 + o(1))d^{\beta-2}\frac{(\beta-2)^{\beta-1}m^{3-\beta}}{(\beta-1)^{\beta-2}(3-\beta)} & \text{if } 2 < \beta < 3. \end{cases} \tag{3}$$

3   Local Graphs and Hybrid Graphs

3.1   Local Graphs

There are a number of ways to define local connectivity between two given vertices u and v. A natural approach is to consider the maximum number $a(u, v)$

of short edge-disjoint paths between the vertices, where *short* means having length at most ℓ. Another approach is to consider the minimum size $c(u, v)$ of a set of edges whose removal leaves no short path between the vertices. When we restrict to short paths, the analogous version of the max-flow min-cut theorem does not hold, and in fact a and c can be different by a factor of $\Theta(\ell)$ ([2]). However we still have the trivial relations $a \leq c \leq \ell \cdot a$.

Both of the above notions of local connectivity are difficult to compute, and in fact computing the maximum number of short disjoint paths is $\mathcal{NP}$-hard if $\ell \geq 4$ [9]. Instead we will consider the maximum short flow between u and v. The maximum short flow can be computed in polynomial time using nontrivial but relatively efficient algorithms for fractional packing (see Section 3.2). The most compelling reason for using short flow as a measure of local connectivity is that it captures the spirit of $a(u, v)$ and $c(u, v)$, but is efficiently computable.

Formally, a short flow is a positive linear combination of short paths where no edge carries more than 1 unit of flow. Finding the maximum short flow can be viewed as a linear program. Let P_ℓ be the collection of short u-v paths, and let P_e be the collection of short u-v paths which intersect the edge e.

Definition 1 (Flow Connectivity). *A short flow is a feasible solution to the following linear program. The flow connectivity $f(u, v)$ between two vertices is the maximum value of any short flow, which is the optimum value of the following LP problem:*

$$\text{maximize} \quad \sum_{p \in P_\ell} f_p \tag{4}$$

$$\text{subject to} \quad \sum_{p \in P_e} f_p \leq 1 \qquad \text{for each } e \in L.$$

$$f_p \geq 0 \qquad \text{for each } p \in P_\ell.$$

The linear programming dual of the flow connectivity problem is a fractional cut problem: to find the minimum weight cutset so every short path has at least 1 unit of cut weight. This gives us the following relation between a, c, and f.

$$a \leq f \leq c \leq \ell \cdot a.$$

We say two vertices u and v are (f, ℓ)-connected if there exists an short flow between them of size at least f.

Definition 2 (Local Graphs). *A graph L is an (f, ℓ)-local graph if for each edge $e = (u, v)$ in L, the vertices u and v are (f, ℓ)-connected in $L \setminus \{e\}$.*

3.2   Computing the Maximum Short Flow

The problem of finding the maximum short flow between u and v can be viewed as a fractional packing problem, as introduced by Plotkin, Shmoys, and Tardos [13]. A fractional packing problem has the form

$$\max\{\, \mathbf{c}^{\mathbf{T}}\mathbf{x} \mid A\mathbf{x} \leq \mathbf{b}, \mathbf{x} \succeq \mathbf{0}\,\}.$$

To view the maximum short flow as a fractional packing problem, first let $G(u,v)$ be a subgraph containing all short paths from u to v. For example, we may take $G(u,v) = N_{\ell/2}(u) \cup N_{\ell/2}(v)$. Let A be the incidence matrix where each row represents an edge in $G(u,v)$ and each column represents a short path from u to v. Let $\mathbf{b} = \mathbf{c} = \mathbf{1}$.

Using the fractional packing algorithm of Garg and Könemann in [8], the maximum short flow can be computed approximately with the following time bounds. See the full version [1] for further details.

Theorem 1. *A $(1-\epsilon)^{-2}$-approximation to the maximum short flow can be computed in time $O(m^2 \ell \lceil \frac{1}{\epsilon} log_{1+\epsilon} m \rceil)$, where m is the number of edges in $G(u,v)$.*

3.3 Hybrid Power Law Graphs

A hybrid graph H is the union of the edge sets of an (f, ℓ)-local graph L and a global graph G on the same vertex set. We here consider the case where the global graph $G(\mathbf{w})$ is a power law graph $M(n, \beta, d, m)$. In this case, the hybrid graph will have small diameter and average distances, due to the following results on random power law graphs [3].

Theorem 2. *For a random power law graph $G = M(n, \beta, d, m)$ and $\beta > 3$, almost surely, the average distance is $(1+o(1))\frac{\log n}{\log d}$ and the diameter is $O(\log n)$.*

Theorem 3. *For a random power law graph $G = M(n, \beta, d, m)$ and $2 < \beta < 3$, almost surely, the average distance is $O(\log \log n)$ and the diameter is $O(\log n)$. For a random power law graph $G = M(n, \beta, d, m, L)$ and $\beta = 3$, almost surely, the average distance is $O(\log n / \log \log n)$ and the diameter is $O(\log n)$.*

The diameter of the hybrid graph can be smaller if the local graph satisfies additional conditions. A local graph L is said to have isoperimetric dimension δ if for every vertex v in L and every integer $k < (\log \log n)^{1/\delta}$, there are at least k^{δ} vertices in L of distance k from v. For example, the grid graph in the plane has isoperimetric dimension 2, and the d-dimensional grid graph has isoperimetric dimension d. The following were proved in [4].

Theorem 4. *In a hybrid graph H with $G = M(n, \beta, d, m, L)$ and $2 < \beta < 3$, suppose that L has isoperimetric dimension $\delta \geq \log \log n / (\log \log \log n)$. Then almost surely, the diameter is $O(\log \log n)$.*

Theorem 5. *In a hybrid graph H with $G = M(n, \beta, d, m, L)$ and $2 < \beta < 3$, suppose that the local graph has isoperimetric dimension δ. Then almost surely, the diameter is $O((\log n)^{1/\delta})$.*

Theorem 6. *In a hybrid graph H with $G = M(n, \beta, d, m, L)$ and $2 < \beta < 3$, suppose that every vertex is within distance $\log \log n$ of some vertex of degree $\log n$. Then almost surely, the diameter is $O(\log \log n)$.*

4   Extracting the Local Graph

For a given graph, the problem of interest is to extract the largest (f, ℓ)-local subgraph. We define $L_{f,\ell}(G)$ to be the union of all (f, ℓ)-local subgraphs in H. By definition, the union of two (f, ℓ)-local graphs is an (f, ℓ)-local graph, and so $L_{f,\ell}(G)$ is in fact the unique largest (f, ℓ)-local subgraph in G. We remark that $L_{f,\ell}(G)$ is not necessarily connected. There is a simple greedy algorithm to compute $L_{f,\ell}(G)$ in any graph G.

4.1   An Algorithm to Extract the Local Graph

Extract(f, ℓ):   We are given as input a graph G and parameters (f, ℓ). Let $H = G$. If there is some edge $e = (u, v)$ in H where u and v are not (f, ℓ)-connected in $H \setminus \{e\}$, then let $H = H \setminus \{e\}$. Repeat until no further edges can be removed, then output H.

Theorem 7. *For any graph G and any (f, ℓ), **Extract**(f, ℓ) returns $L_{f,\ell}(G)$.*

Proof. Given a graph G, let L' be the graph output by the greedy algorithm. A simple induction argument shows that each edge removed by the algorithm is not part of any (f, l)-local subgraph of G, and thus $L_{f,\ell}(G) \subseteq L'$. Since no further edges can be removed from L', L' is (f, l)-local and so $L' \subseteq L_{f,\ell}(G)$. Thus $L' = L_{f,\ell}(G)$.

The algorithm requires $O(|E|^2)$ maximum short flow computations.

4.2   Recovering the Local Graph

When applied to a hybrid graph $H = G \cup L$ with an (f, ℓ)-local graph L, the algorithm **Extract**(f, ℓ) will output $L_{f,\ell}$, which is almost exactly L if G is sufficiently sparse. Note that $L \subseteq L_{f,\ell}$ by definition of the local graph. The proof of Theorem 8 is outlined in section 5. For complete proofs of the following theorems, see the full version [1].

Theorem 8. *Let $H = G \cup L$ be a hybrid graph where L is (f, ℓ)-local with maximum degree M, and where $G = G(w)$ with average weight d, second order average weight $\tilde{d}$, and maximum weight m. Let $L' = L_{f,\ell}(H)$. If $\tilde{d}$ satisfies*

$$\tilde{d} \leq n^\alpha \leq \left(\frac{nd}{m^2}\right)^{1/\ell} n^{-3/f\ell} \text{ for some constant } \alpha > 0,$$

Then with probability $1 - O(n^{-1})$:

1. *$L' \setminus L$ contains $O(\tilde{d})$ edges.*
2. *$d_{L'}(x, y) \geq \frac{1}{\ell} d_L(x, y)$ for every pair $x, y \in L$.*

In the special case where all the weights are equal and $G(w) \sim G(n, p)$, Theorem 8 has a cleaner statement.

Theorem 9. *Let H be a hybrid graph as in Theorem 8 and let $G = G(n, p)$ with $p = dn^{-1}$. If*

$$d \leq n^{\alpha} \leq n^{1/\ell} n^{-3/f\ell} \text{ for some constant } \alpha > 0,$$

Then with probability $1 - O(n^{-1})$, results (1)-(2) from Theorem 8 hold.

This result is tight in the sense that if d is larger than $n^{\frac{1}{\ell}}$ we cannot hope to recover a good approximation to the original local graph.

Theorem 10. *Let G be chosen from $G(n, p)$ where $p = dn^{-1}$, and let*

$$d \geq 6fn^{\frac{1}{\ell}} (\log n)^{\frac{1}{\ell}}.$$

With probability $1 - O(n^{-2})$, $L_{f,\ell}(H) = H$.

We also point out that the term $\left(\frac{nd}{m^2}\right)^{1/\ell}$ in Theorem 8 is nearly optimal, although we will not make this precise. In the $G(\mathbf{w})$ model, $\tilde{d}$ is roughly the factor we expect a small neighborhood to expand at each step, and dependence on m is unavoidable for a certain range of β.

4.3   Experiment

We have implemented the **Extract** algorithm and tested it on various graphs. For some hybrid graphs, the local graphs are almost perfectly recovered (see figure 2).

4.4   Communities and Examples

The local graph L found by the **Extract**(f, l) algorithm is not necessarily connected. Each connected component of L can be viewed as a local community. By fixing l and increasing f we obtain a hierarchy of successively smaller communities.

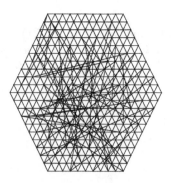
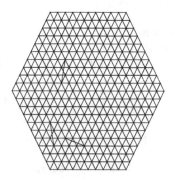

Fig. 1. *A hybrid graph, containing the hexagonal grid graph as a local graph*

Fig. 2. *After applying **Extract** (with parameters $k = 2.5$ and $l = 4$), the local graph is almost perfectly recovered*

Flake et al. [7] defined a hierarchy of communities using minimum cut trees. Their communities have provably good expansion, and few edges between communities. The communities found by **Extract** are highly locally connected, and are robust in the sense of Theorem 8. These communities can have rich structures other than cliques or complete bipartite subgraphs. The communities are monotone in the sense that adding edges only increases the size of a community.

We applied the **Extract** algorithm to a routing graph G collected by "champagne.sdsc.org". The maximum 3-connected subgraph of G consists of 7 copies of K_4 and a large connected component L with 2364 vertices and 5947 edges. The **Extract**$(3,3)$ algorithm breaks L into 79 non-trivial communities. The largest community has 881 vertices. The second largest community (of size 59) is illustrated in Figure 3, and two communities of size 25 and 35 are illustrated in Figures 5 and 6.

5   Proof of Theorem 8

We say an edge in H is *global* if it is in $G \setminus L$. A global edge is *long* if $d_L(u,v) > \ell$ and *short* otherwise. We will show that under the hypotheses of Theorem 8 no

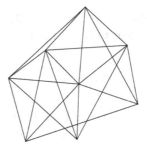

Fig. 3. *A (3,3)-connected community of size 59 in a routing graph*

Fig. 4. *A (4,3)-connected sub-community of the community in Figure 5*

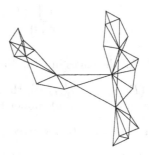

Fig. 5. *A (3,3)-connected community of size 35 in the routing graph*

Fig. 6. *A (3,3)-connected community of size 25 in the routing graph*

long edges are likely to survive. If (u, v) is a short edge in G, it is possible that there is a short flow of size f from u to v entirely through edges in L. This means we can not say a short edge is unlikely to survive without placing additional assumptions on the local graph. However, an easy computation shows there are not likely to be many short edges in G.

Proposition 1. *The expected number of short edges in $L_{f,\ell}(G) \setminus L$ is $O(\tilde{d})$.*

The proof is included in the full version. In the rest of this section we outline the proof of the following.

Proposition 2. *The probability that a given long edge survives is $O(n^{-3})$.*

Definition 3. $N_k(u)$ and $\Gamma_k(u)$.

For $k \in [0, \ell]$, let $N_k(u)$ be the set of vertices y such that there exists a path $p = v_0 \ldots v_k$ from u to y in H obeying the following condition:

$$d_L(v_i, v) > \ell - i \text{ for all } i \in [0, k].$$

We define $\Gamma_k(u)$ to be the corresponding strict neighborhood,

$$\Gamma_k(u) = \{ y \mid y \in N_k(u), y \notin N_0(u) \ldots N_{k-1}(u) \}.$$

The following recursive definition of $\Gamma_k(u)$ will be useful, and is easily seen to be equivalent to the original.

$$\Gamma_0(u) = \{u\}.$$
$$\Gamma_k(u) = \left\{ y \mid \begin{array}{l} y \notin N_{\ell-k}(v), \\ y \notin \Gamma_0(u) \ldots \Gamma_{k-1}(u), \\ (x, y) \in H \text{ for some } x \in \Gamma_{k-1}(u). \end{array} \right\}$$

Definition 4. $C(u, v)$.

Define $C(u, v)$ to be the set of edges in

$$\bigcup_{k \in [1, \ell]} \left(\Gamma_{k-1}(u) \times N^L_{\ell-k}(v) \right).$$

Remark 1. All the edges in C(u,v) are global edges.

If $(x, y) \in \left(\Gamma_{k-1}(u) \times N^L_{\ell-k}(v) \right)$, then $d_L(x, v) > \ell - (k - 1)$ and $d_L(y, v) \leq \ell - k$. Thus $d_L(x, y) \geq 2$, so (x, y) cannot be a local edge and must be global.

Lemma 1. *Let (u, v) be a surviving long edge in H. Then $f \leq |C(u, v)|$.*

Proof. We first show that every short path between u and v in H contains an edge from $C(u, v)$. Let $p = v_0 \ldots v_k$ be a path of length $k \leq \ell$ between u and v in H. The last vertex on the path is $v_k = v$, so we have $d_L(v_k, v) = 0 \leq \ell - k$, and thus

$v_k \notin N_k(u)$. The first vertex on the path is $v_0 = u$, and thus $v_0 \in N_0(u)$. Let $j \geq 1$ be the smallest integer such that $v_j \notin N_j(u)$. By definition, $v_{j-1} \in N_{j-1}(u)$, while $v_j \notin N_j(u)$. This implies $d_L(v_j, v) \leq \ell - j$, so $v_j \in N_{\ell-j}^L(v)$. We now have that $v_{j-1}v_j$ is an edge in $N_{j-1}(u) \times N_{\ell-j}^L(v)$. We conclude that $v_{j-1}v_j$ is an edge in $C(u,v)$ by noticing

$$N_{j-1}(u) \times N_{\ell-j}^L(v) \subseteq \bigcup_{k \in [1,j]} \left(\Gamma_{k-1}(u) \times N_{\ell-k}^L(v) \right) .$$

We can now complete the proof. If the set $C(u,v)$ is removed, then no short paths remain between u and v. Thus, if a is the maximum number of short disjoint paths, c is the size of the minimum cut to remove all short paths, and f is the maximum ℓ-flow, we have

$$a \leq f \leq c \leq |C(u,v)| .$$

Thus $f \leq |C(u,v)|$, and in fact the lemma would also hold if we were considering disjoint paths or cuts.

Lemma 2. *If we condition on the values of the sets $\Gamma_0(u) \ldots \Gamma_{\ell-1}(u)$, then the edges in*

$$\bigcup_{k \in [1,\ell]} \left(\Gamma_{k-1}(u) \times N_{\ell-k}^L(v) \right)$$

are mutually independent and occur with the same probabilities as in G.

Proof. We will reveal $\Gamma_0(u) \ldots \Gamma_{\ell-1}(u)$ sequentially by a breadth-first search. From the recursive definition of $\Gamma_k(u)$, it is clear that we can determine $\Gamma_k(u)$ given $\Gamma_{k-1}(u)$ by examining only the edges in

$$\Gamma_{k-1} \times \left(V \setminus (N_{\ell-k}^L \cup \Gamma_0(u) \cup \cdots \cup \Gamma_{k-1}(u)) \right) .$$

In particular, in determining $\Gamma_k(u)$ from $\Gamma_{k-1}(u)$ we do not examine any edges with an endpoint in $\Gamma_0(u) \ldots \Gamma_{k-2}(u)$, and we do not examine any edges in $\Gamma_{k-1}(u) \times N_{\ell-k}^L(v)$. Thus, we do not examine any edges in

$$\bigcup_{j \in [1,k]} \left(\Gamma_{k-1}(u) \times N_{\ell-k}^L(v) \right)$$

when revealing $\Gamma_k(u)$, and we may reveal $\Gamma_0(u) \ldots \Gamma_{\ell-1}(u)$ without examining any edges in

$$\bigcup_{k \in [1,\ell]} \left(\Gamma_{k-1}(u) \times N_{\ell-k}^L(v) \right) .$$

Lemma 3. *With probability $1 - e^{-\Omega(n^\alpha)}$,*

$$\sum_{k \in [1,\ell]} Vol(\Gamma_{k-1}(u)) Vol(N_{\ell-k}^L(v)) \leq 4m^2 (4M\hat{d})^{\ell-1} .$$

Proof. Included in the full version [1]

Proof of Proposition 2: Let

$$E = \bigcup_{k \in [1,\ell]} \left(\Gamma_{k-1}(u) \times N_{\ell-k}^L(v) \right) .$$

$|C(u,v)|$ is the number of global edges in E. If (u,v) is a surviving long edge, then $|C(u,v)| \geq f$ by Lemma 1. Let E^f denote the set of ordered f-tuples from E with distinct entries. Let B be the event that

$$\sum_{k \in [1,\ell]} Vol(\Gamma_{k-1}(u))Vol(N_{\ell-k}^L(v)) \leq 4m^2(4M\hat{d})^{\ell-1} ,$$

which occurs with probability $1 - e^{-\Omega(n^\alpha)}$ by Lemma 3.

$$\Pr\left[\,|C(u,v)| \geq f\,\right] \leq \Pr\left[\,|C(u,v)| \geq f \mid B\,\right] + \Pr\left[\,|C(u,v)| \geq f \mid \bar{B}\,\right]$$
$$\leq \Pr\left[\,|C(u,v)| \geq f \mid B\,\right] + e^{-\Omega(n^\alpha)} .$$

We will now bound $\Pr\left[\,|C(u,v)| \geq f \mid B\,\right]$. We will first determine E by revealing the sets $\Gamma_0(u) \ldots \Gamma_{\ell-1}(u)$. Critically, Lemma 2 tells us that the potential edges in E are mutually independent and occur with the same probabilities as in G. Thus,

$$\Pr\left[\,|C(u,v)| \geq f \mid B\,\right] \leq \sum_{((x_1,y_1)\ldots(x_f,y_f)) \in E^f} \Pr\left[\bigwedge_{i \in [1,f]} (x_i, y_i) \in G\right]$$

$$= \sum_{((x_1,y_1)\ldots(x_f,y_f)) \in E^f} \prod_{i \in [1,f]} w_{x_i} w_{y_i} \rho$$

$$\leq \rho^f \left(\sum_{k \in [1,\ell]} Vol(\Gamma_{k-1}(u))Vol(N_{\ell-k}^L(v)) \right)^f$$

$$\leq \rho^f \left(4m^2(4M\hat{d})^{\ell-1} \right)^f$$

$$= \left(4m^2(4M\hat{d})^{\ell-1}\rho \right)^f .$$

Since $\hat{d} = n^\alpha \leq (\frac{nd}{m^2})^{1/\ell}n^{-3/f\ell}$,

$$\Pr[|C(u,v)| \geq f \mid B] \leq \left(4m^2(4Mn^\alpha)^{\ell-1}\rho\right)^f \leq \left(4m^2(4M)^{\ell-1}(n^\alpha)^\ell \frac{1}{nd}\right)^f$$

$$\leq \left((4M)^\ell n^{-3/f} \right)^f$$

$$= O(n^{-3}) .$$

Thus the probability that a given long edge survives is at most

$$O(n^{-3}) + e^{-\Omega(n^\alpha)} = O(n^{-3}) .$$

Proof of Theorem 8: Since there are at most n^2 edges in G, with probability $1 - O(n^{-1})$ no long edges survive. In that case, $L' \setminus L$ contains only short edges, and there are $O(\bar{d})$ of these by Proposition 1, so part (1) follows. To prove (2), note that if no long edges survive, then all edges in L' must be short. If (u, v) is an edge in L', $d_L(u, v) \leq \ell$. If p' is a path between two vertices x, y in L' with length k, then by replacing each edge with a short path we obtain a path p in L between x and y with length at most ℓk. The result follows.

References

1. R. Andersen, F. Chung, L. Lu, Modelling the small world phenomenon using local network flow (Preprint)
2. S. Boyles, G. Exoo, On line disjoint paths of bounded length. *Discrete Math.* **44** (1983)
3. F. Chung and L. Lu, Average distances in random graphs with given expected degree sequences, *Proceedings of National Academy of Science*, **99** (2002).
4. F. Chung and L. Lu, The small world phenomenon in hybrid power law graphs *Lecture Note in Physics* special volume on "Complex Network", to appear.
5. P. Erdős and T. Gallai, Gráfok előírt fokú pontokkal (Graphs with points of pre-scribed degrees, in Hungarian), *Mat. Lapok* **11** (1961), 264-274.
6. A. Fabrikant, E. Koutsoupias and C. H. Papadimitriou, Heuristically optimized trade-offs: a new paradigm for power laws in the Internet, *STOC* 2002.
7. G. W. Flake, R. E. Tarjan, and K. Tsioutsiouliklis, Graph Clustering and Minimum Cut Trees.
8. N. Garg, J. Konemann, Faster and simpler algorithms for multicommodity flow and other fractional packing problems. *Technical Report, Max-Planck-Institut fur Informatik, Saarbrucken, Germany* (1997).
9. A. Itai, Y. Perl, and Y. Shiloach, The complexity of finding maximum disjoint paths with length constraints, *Networks* **12** (1982)
10. J. Kleinberg, The small-world phenomenon: An algorithmic perspective, *Proc. 32nd ACM Symposium on Theory of Computing*, 2000.
11. S. Milgram, The small world problem, *Psychology Today*, **2** (1967), 60-67.
12. M. Mitzenmacher, A Brief History of Generative Models for Power Law and Log-normal Distributions, *Internet Math.* 1 (2003), no. 2.
13. S. Plotkin, D. B. Shmoys, and E Tardos, Fast approximation algorithms for fractional packing and covering problems, *FOCS* 1991, pp. 495–504.
14. D. J. Watts and S. H. Strogatz, Collective dynamics of 'small world' networks, *Nature* **393**, 440-442.

Dominating Sets in Web Graphs[*]

Colin Cooper[1], Ralf Klasing[2,**], and Michele Zito[3]

[1] Department of Computer Science, King's College, London WC2R 2LS (UK)
ccooper@dcs.kcl.ac.uk
[2] MASCOTTE project, I3S-CNRS/INRIA, Université de Nice-Sophia Antipolis,
2004 Route des Lucioles, BP 93, F-06902 Sophia Antipolis Cedex (France)
Ralf.Klasing@sophia.inria.fr
[3] Department of Computer Science, University of Liverpool,
Peach Street, Liverpool L69 7ZF (UK)
M.Zito@csc.liv.ac.uk

Abstract. In this paper we study the size of generalised dominating sets in two graph processes which are widely used to model aspects of the world-wide web. On the one hand, we show that graphs generated this way have fairly large dominating sets (i.e. linear in the size of the graph). On the other hand, we present efficient strategies to construct small dominating sets.

The algorithmic results represent an application of a particular analysis technique which can be used to characterise the asymptotic behaviour of a number of dynamic processes related to the web.

1 Introduction

In recent years the world wide web has grown dramatically. Its current size is measured in billions of pages [20], and pages are added to it every day. As this graph (nodes correspond to web pages and edges to links between pages) continues to grow it becomes increasingly important to study mathematical models which capture its structural properties [6, 19]. Such models can be used to design efficient algorithms for web applications and may even uncover unforeseen properties of this huge evolving structure. Several mathematical models for analysing the web have been proposed (for instance [5, 9, 19]). The (evolution of the) web graph is usually modelled by a (random) process in which new vertices appear from time to time. Such vertices may be linked randomly to the existing structure through some form of *preferential attachment*: existing vertices with many neighbours are somewhat more likely to be linked to the newcomers.

The main focus of research so far has been on capturing empirically observed [3, 6] features of the web. No attempt has been made to characterise graph-theoretic sub-structures of such graphs. We initiate such investigation by looking

---

[*] The first two authors were partially supported by a Royal Society Grant under the European Science Exchange Programme.

[**] Partially supported by the European projects RTN ARACNE (contract no. HPRN-CT-1999-00112) and IST FET CRESCCO (contract no. IST-2001-33135). Partial support by the French CNRS AS Dynamo.

S. Leonardi (Ed.): WAW 2004, LNCS 3243, pp. 31–43, 2004.

at sets of vertices that, in a sense, cover all other vertices. More formally, a vertex in a graph *dominates* all vertices that are adjacent to it (parallel edges give multiple domination). In the spirit of Harary and Haynes [14], an *h-dominating set* for a graph $G = (V, E)$ is a set $S \subseteq V$ such that each vertex in $V \setminus S$ is dominated at least h times by vertices in S. Let $\gamma_h = \gamma_h(G)$ denote the size of the smallest h-dominating sets in G. The *minimum h-dominating set* problem (MhDS) asks for an h-dominating set of size γ_h.

Dominating sets play an important role in many practical applications, e.g. in the context of distributed computing or mobile ad-hoc networks [2, 10, 22]. The reader is referred to [15, 16] for an in-depth view of the subject. The typical fundamental task in such applications is to select a subset of nodes in the network that will 'provide' a certain service to all other vertices. For this to be time-efficient, all other vertices must be directly connected to the selected nodes, and in order for it to be cost-effective, the number of selected nodes must be minimal. In relation to web graphs, a dominating set may be used to devise efficient web searches. For $h > 1$ an h-dominating set can be considered as a more fault-tolerant structure. If up to $h - 1$ vertices or edges fail, the domination property is still maintained (i.e. it is still possible to provide the service).

The MhDS problem is NP-hard [13, 18] and, moreover, it is not likely that it may be approximated effectively [12, 18]. Polynomial time algorithms exist on special classes of graphs (e.g. [21]). The M1DS problem has been studied also in random graphs. In the *binomial model* $G(n, p)$ [24] the value of γ_1 can be pin-pointed quite precisely, provided p is not too small compared to n. In random regular graphs of degree r (see for example results in the *configuration model* [26] and references therein) upper and lower bounds are known on γ_1.

In this paper we look at simple and efficient algorithms for building small h-dominating sets in graphs. The performance guarantees of these algorithms are analysed under the assumption that the input is a random web graph. We also analyse the tightness of the performances of such algorithms, by proving combinatorial lower bounds on γ_h, for any fixed $h \geq 1$. Such bounds, often disappointingly weak, offer nevertheless a proof that, most of the times, the sets returned by the various algorithms are only a constant factor away from an optimal answer. Finally, we compare the quality of the solutions returned by (some of) our algorithms with empirical average values of γ_1.

The main outcome of this paper can be stated informally by saying that web graphs have fairly large dominating sets. Hence a crawler who wants to use a dominating set to explore the web will need to store a large proportion of the whole graph. Interestingly, the results in this paper also uncover a(nother) difference between models of the web based on preferential attachment and more traditional random graph models. The tendency to choose neighbours of high degree affects the size of the smallest dominating sets.

Most of our algorithms are *on-line* in the sense that the decision to add a particular vertex to the dominating set is taken without total information about the web graph under consideration, and *greedy* in the sense that decisions, once taken, are never changed. The algorithms are also quite efficient: only a constant amount of time is used per update of the dominating set. Such algorithms are of particular interest in the context of web graphs. As the web graph is evolving, one wants to decide whether a new vertex is to be added to the already existing dominating set without recomputing the existing dominating set and with minimal computational effort. On-line strategies for the dominating set problem have been considered in the past [11, 17] for general graphs. However the authors are not aware of any result on on-line algorithms for this problem in random graphs. Our results hold *asymptotically almost surely* (a.a.s.), i.e. with probability approaching one as the size of the web graph grows to infinity. The algorithmic results are based on the analysis of a number of (Markovian) random processes. In each case the properties of the process under consideration lead to the definition of a (deterministic) continuous function that is very close (in probability) to the values of the process, as the size of the graph grows. It should be pointed out at this stage that the proposed analysis methodology is quite general. We apply it to analyse heuristics for the MhDS problem only, but it would allow to prove results about other graphs parameters such as the independence or the chromatic number. The method is closely related to the so called *differential equation method* [25]. In fact a version of the main analytical tool proposed by Wormald can be adapted to work for the processes considered in this paper. However the machinery employed in [25] is not needed to analyse the processes considered in this paper. Our results are obtained by proving concentration of the various processes of interest around their mean and by devising a method for getting close estimates on the relevant expectations. In Section 2 we review the definitions of the models of web graphs that we will use. We also state our main result in the context of these models, and present more detailed comments on our general analysis method. In the following section we consider a very simple algorithm and apply the proposed method to obtain non-trivial upper bounds on γ_1. Refined algorithms are introduced and analysed in Section 4 and 5. In Section 6 we discuss generalisations to $h > 1$. Then we turn to lower bounds. In Section 7 and 8 we present our argument for the lower bounds stated in Section 2. Finally we briefly comment on some empirical work carried out on a sub-class of the graphs considered in this paper.

2   Models and Results

The models used in this paper are based on the work of Albert and Barabasi [3]. A *web graph* (see also [9]) can be defined as an ever growing structure in which, at each step, new vertices, new edges or a combination of these can be added. Decisions on what to add to the existing graph are made at random based on the values of a number of defining parameters. The existence of these parameters makes the model very general. For the purposes of this paper, to avoid cluttering the description of our results, we prefer to make a drastic simplification. We will consider graphs generated according to two rather extreme procedures derived from the general model. In each case the generation process is governed by a single integer parameter m. The second of these mimics the prefential attachment phenomenon. The first one, related to more traditional random graph models, is considered mainly for comparison.

Random Graph. The initial graph $G_0^{R,m}$ is a single vertex v_0 with m loops attached to it. For $t \geq 1$, let $G_{t-1}^{R,m}$ be the graph generated in the first $t-1$ steps of this process, to define $G_t^{R,m}$ a new vertex v_t is generated, and it is connected to $G_{t-1}^{R,m}$ through m (undirected) edges. The neighbours of v_t are chosen uniformly at random (u.a.r.) with replacement from $\{v_0, \ldots, v_{t-1}\}$.

Pure Copy. The initial graph $G_0^{C,m}$ is a single vertex v_0 with m loops attached to it. For $t \geq 1$, graph $G_t^{C,m}$ is defined from $G_{t-1}^{C,m}$ by generating a new vertex v_t, and connecting it to $G_{t-1}^{C,m}$ through m (undirected) edges. A vertex $u \in \{v_0, \ldots, v_{t-1}\}$ is connected to v_t with probability $\frac{|\Gamma(u)|}{2mt}$ (where $\Gamma(u) = \{w : \{u, w\} \in E(G_{t-1}^{C,m})\}$).

Both models are dynamic, with new vertices and edges continuously increasing the size of the graph. However they represent two extreme cases. In the random graph model connections are completely random, whereas in the copy model they are chosen based on the current popularity of the various vertices. The results in this paper are summarised by the following Theorem.

Theorem 1. *Let $M \in \{R, C\}$, and m be a positive integer. Then for each $h \geq 1$ there exist positive real constants α_{lo}^M and α_{up}^M (dependent on M, m and h but independent of t) with $\alpha_{lo}^M < \alpha_{up}^M < 1$ such that $\alpha_{lo}^M \cdot t \leq \gamma_h(G_t^{M,m}) \leq \alpha_{up}^M \cdot t$ a.a.s.*

For $h = 1$, the values of the constants mentioned in the Theorem are reported in Table 1. Bounds for $h > 1$ are reported (and briefly commented) in Section 6. In particular the upper bounds are proved by analysing the size of the dominating set returned by a number of simple polynomial time algorithms.

The proof of Theorem 1 is based on the fact that natural edge-exposure martingales can be defined on the graph processes under consideration [9]. More precisely, if $f(G)$ is any graph theoretic function (e.g. the size of the dominating set returned by a particular algorithm), the random process defined by setting $Z_0 = E(f(G_t^{M,m}))$, and Z_i (for $i \in \{1, \ldots, mt\}$) to be the expectation of $f(G_t^{M,m})$

Table 1. Numerical values defined in Theorem 1 for the minimum dominating set problem

m	α_{lo}^R	α_{up}^R	α_{lo}^C	α_{up}^C
1	0.3678	0.5	0.21	0.3333
2	0.0585	0.3714	0.0286	0.2342
3	0.0422	0.3054	0.0148	0.1777
4	0.0352	0.2634	0.0097	0.1422
5	0.0261	0.2335	0.0066	0.1178
6	0.0247	0.2110	0.0049	0.1000
7	0.0208	0.1932	0.0038	0.0865

conditioned on the "exposure" of the first i edges in the graph process is a martingale. Notice that the space of all graphs which can be generated according to the given model $G_t^{M,m}$ is partitioned into classes (or i-blocks) containing all those graphs which coincide w.r.t. the first i edge exposures.

In the forthcoming sections we will repeatedly use the following concentration result (for a proof see, for instance, [1]).

Theorem 2. *Let* $c = Z_0, \ldots, Z_n$ *be a martingale with* $|Z_{i+1} - Z_i| \leq 1$ *for all* $i \in \{0, \ldots, n-1\}$. *Then* $\Pr[|Z_n - c| > \lambda\sqrt{n}] \leq 2e^{-\lambda^2/2}$.

In all our applications $c = \mathrm{E}(f(G_t^{M,m}))$, $n = mt$ and $\lambda = O(\log t)$. In order to apply Theorem 2 one needs to prove that $|Z_{i+1} - Z_i| \leq 1$. Such inequality follows from the smoothness of f (i.e. $|f(G) - f(H)| \leq 1$ if G and H differ w.r.t. the presence of a single edge) and the ability to demonstrate the existence of a measure preserving bijection between $i + 1$-blocks in a same i-block. This is obvious in the random graph process as edges are inserted independently. In the case of the copy model it is convenient to identify each graph with the projection of a particular type of *configuration*, i.e. an ordered collection of mt labelled pairs of points (see [4]). Let C_1 and C_2 be two such configurations that are identical up to the ith pair. Suppose the $i + 1$-st pair is $\{a, b\}$ in C_1 and $\{a, c\}$ in C_2. If C_1 never uses point b again then the image of C_1 under the measure preserving bijection will be a configuration C' identical to C_1 except that pair $\{a, b\}$ is replaced by pair $\{a, c\}$. If b is used in a following pair (say $\{d, b\}$) of C_1 then C' will have $\{a, c\}$ instead of $\{a, b\}$ and $\{d, c\}$ instead of $\{d, b\}$, and so on. Similar construction is presented in [9].

Finally we will need the following.

Lemma 1. *If* Z_n *is a martingale and there exist constants* $c_1, c_2 > 0$ *such that* $\mu_n = \mathrm{E}(Z_n) \in [c_1 n, c_2 n]$, *then for each fixed integer* $j > 0$, *there exists positive constant* K *and* ϵ *such that* $|(Z_n)^j - (\mu_n)^j| \leq K n^{j-\epsilon}$ *a.a.s..*

Proof. If Z_n is a martingale, then by Theorem 2, $\mu_n - \lambda\sqrt{n} \leq Z_n \leq \mu_n + \lambda\sqrt{n}$ where w.l.o.g. we assume $\lambda = o(\sqrt{n})$. From this we also have, for each fixed integer $j > 0$,

$$(\mu_n)^j\left(1 - \frac{\lambda\sqrt{n}}{\mu_n}\right)^j \leq (Z_n)^j \leq (\mu_n)^j\left(1 + \frac{\lambda\sqrt{n}}{\mu_n}\right)^j.$$

The result now follows (provided K is chosen big enough) since, the assumptions on λ and μ_n entail that $\left(1 + \frac{\lambda\sqrt{n}}{\mu_n}\right)^j$ is at most $1 + \frac{j^2\lambda\sqrt{n}}{\mu_n}$, whereas $\left(1 - \frac{\lambda\sqrt{n}}{\mu_n}\right)^j$ is at least $1 - \frac{2j\lambda\sqrt{n}}{\mu_n}$. □

3   Simplest Algorithm

The algorithm presented in this section is a very simple "first attempt" solution for the problem at hand. Although in many cases it does not lead to a very small dominating set, it represents a natural benchmark for any more refined heuristic.

Algorithm 1. Before the first step of the algorithm the graph consists of a single vertex v_0 and $S = \{v_0\}$. At step t if the newly generated vertex v_t does not have any neighbours in S (i.e. $\Gamma(v_t) \cap S = \emptyset$) then v_t is added to S.

In the forthcoming discussion m is a fixed positive integer. Let X_t denote the size of the dominating set S computed by Algorithm 1 before v_t is added to the current graph and let $\mu_t = E(X_t)$. For graphs generated according to the $G_t^{R,m}$ model, the probability that v_t misses the dominating set is $(1 - \frac{X_t}{t})^m$. Hence we can write

$$\mu_{t+1} = \mu_t + E[(1 - \frac{X_t}{t})^m].$$

Let $x = x(m)$ be the unique solution of the equation $x = (1 - x)^m$ in $(0,1)$. Table 2 gives the values of x for the first few values of m.

Table 2. Numerical values defined in Lemma 2

m	x
1	0.5
2	0.382
3	0.3177
4	0.2755
5	0.2451
6	0.2219
7	0.2035

Lemma 2. *For any $\frac{1}{2} < \rho < 1$ constant, there exists an absolute positive constant C such that for all $t > 0$, $|\mu_t - xt| \le Ct^\rho$ a.a.s.*

Proof. We claim that the difference $|\mu_t - xt|$ satisfies a recurrence of the form

$$|\mu_{t+1} - x(t+1)| \le |\mu_t - xt| + O\left(\sqrt{\frac{\log t}{t}}\right).$$

This can be proved by induction on t. By definition $X_1 = 1$, hence $|\mu_1 - x| = 1 - x$. We also have $|\mu_{t+1} - x(t+1)| = |\mu_t - xt + E[(1 - \frac{X_t}{t})^m] - (1 - x)^m|$. The difference $E[(1 - \frac{X_t}{t})^m] - (1 - x)^m$ can be rewritten as $-\frac{m}{t}(\mu_t - xt) + \{E[(1 - \frac{X_t}{t})^m] - 1 + m\frac{\mu_t}{t} - (1 - x)^m + 1 - mx\}$. Hence

$$|\mu_{t+1} - x(t+1)| = |(1 - \frac{m}{t})(\mu_t - xt) + \{E[(1 - \frac{X_t}{t})^m] - 1 + m\frac{\mu_t}{t} - (1 - x)^m + 1 - mx\}|.$$

To complete the proof notice that, by Lemma 1,

$$E[(1 - \frac{X_t}{t})^m] - 1 + m\frac{\mu_t}{t} = (1 - \frac{\mu_t}{t})^m - 1 + m\frac{\mu_t}{t} + O(\sqrt{\frac{\log t}{t}}).$$

and the function $f(z) = (1 - z)^m - 1 + mz$ satisfies $|f(z_1) - f(z_2)| \le m|z_1 - z_2|$, for $z_1, z_2 \in [0, 1]$. □

The following Theorem is a direct consequence of Lemma 2 and the concentration result mentioned in the previous section.

Theorem 3. $|X_t - xt| = o(t)$ a.a.s.

4   Improved Approximations in the Random Graph Process

Although Algorithm 1 is quite simple, it seems difficult to beat, as a glance at α_{up}^R in Tables 1 and 2 shows. This is especially true for $m = 1$ where no improvement could be obtained. For larger values of m, a better way of finding small dominating sets is obtained by occasionally allowing vertices to be dropped from $\mathcal{S}$. It is convenient to classify the vertices in the dominating set as *permanent* (set $\mathcal{P}$) and *replaceable* (set $\mathcal{R}$). Thus $\mathcal{S} = \mathcal{P} \cup \mathcal{R}$. Let P_t and R_t denote the sizes of such sets at time t (set $P_1 = 0$ and $R_1 = 1$).

Algorithm 2. Before the first step of the algorithm the graph consists of a single vertex v_0 and $\mathcal{R} = \{v_0\}$. After v_t is created and connected to m neighbours, if $\Gamma(v_t) \cap \mathcal{P} \neq \emptyset$ then v_t is moved to $V \setminus \mathcal{S}$. Otherwise v_t is added to $\mathcal{R}$ if $\Gamma(v_t) \cap \mathcal{R} = \emptyset$, otherwise v_t is added to $\mathcal{P}$ and all vertices in $\Gamma(v_t) \cap \mathcal{R}$ are moved to $V \setminus \mathcal{S}$.

The expectations $\pi_t = E(P_t)$ and $\rho_t = E(R_t)$ satisfy:

$$\pi_{t+1} = \pi_t + E[(1 - \tfrac{P_t}{t})^m] - E[(1 - \tfrac{P_t}{t} - \tfrac{R_t}{t})^m],$$
$$\rho_{t+1} = \rho_t + E[(1 - \tfrac{R_t}{t} - \tfrac{P_t}{t})^m] - mE[\tfrac{R_t}{t}(1 - \tfrac{P_t}{t})^{m-1}].$$

Define α_{up}^R as $p + r$, where $p = p(m)$ and $r = r(m)$ satisfy

$$r = \frac{(1-p)^m - p}{1 + m(1-p)^{m-1}},$$
$$p = (1 - p)^m - (1 - p - r)^m.$$

Lemma 3. *For any $\frac{1}{2} < \rho < 1$ constant, there exist absolute positive constants C_1 and C_2 such that for all $t > 0$, $|\pi_t - pt| \leq C_1 t^\rho$ and $|\rho_t - rt| \leq C_2 t^\rho$ a.a.s.*

Proof. The proof is, essentially, a generalisation of that of Lemma 2. We present the argument in some details for π_t. At the inductive step:

$$|\pi_{t+1} - p(t+1)| = |\pi_t - pt + E[(1 - \tfrac{P_t}{t})^m] - (1-p)^m - \{E[(1 - \tfrac{P_t}{t} + \tfrac{R_t}{t})^m] - (1 - p - r)^m\}|.$$

The proof is completed by decomposing the differences $E[(1 - \tfrac{P_t}{t})^m] - (1 - p)^m$ and $E[(1 - \tfrac{P_t}{t} - \tfrac{R_t}{t})^m] - (1 - p - r)^m$ in parts that are proportional to either $P_t - \pi_t$ or $\pi_t - pt$.

The result about ρ_t is proved similarly after noticing that r satisfies:

$$r = (1 - p - r)^m - mr(1 - p)^{m-1}.$$

$\square$

Theorem 2 implies that the sum $p + r$ is in fact very close to $\frac{|S|}{t} = \frac{P_t}{t} + \frac{R_t}{t}$. The values of $p + r$ for $m \leq 7$ are reported in the column labelled α_{up}^R of Table 1.

5   Improved Approximations in the Pure Copy Process

Algorithm 1, described in Section 3, can be analysed in the copy model as well. The expected change in the variable X_t can be computed by keeping track of the total degree of the dominating set, D_t. In particular the following relationships hold

$$E(X_{t+1}) = E(X_t) + E[(1 - \tfrac{D_t}{2mt})^m],$$
$$E(D_{t+1}) = (1 + \tfrac{1}{2t})E(D_t) + mE[(1 - \tfrac{D_t}{2mt})^m].$$

Not surprisingly an analysis similar to the one described in the previous sections implies that such algorithm returns dominating sets of size xt in $G_t^{C,m}$. However, in the copy model we can improve on this by pushing high degree vertices in the dominating set. The natural way to accomplish this would be, for any newly generated uncovered vertex, to select a neighbour of maximum degree and add it to $\mathcal{S}$. Unfortunately such algorithm is not easy to analyse because in the graph processes that we consider there may be few vertices of rather large degree (this is a consequence of the power law distribution of vertex degrees [3, 8]). However a draconian version of this heuristic can be analysed. The following algorithm takes as input an additional integer parameter $k > 0$.

Algorithm 3. Before the first step of the algorithm the graph consists of a single isolated vertex v_0 and $\mathcal{S} = \{v_0\}$. After v_t is created and connected to m neighbours, let Z be the set of all neighbours of v_t in $V \setminus \mathcal{S}$ of degree $km + 1$. If $Z \neq \emptyset$ then all vertices in Z are added to $\mathcal{S}$. Otherwise if v_t is not dominated by some element of $\mathcal{S}$ then a vertex of maximum degree in $\Gamma(v_t)$ is added to the dominating set.

Notice that after each v_t is generated and connected to $G_{t-1}^{C,m}$, all vertices whose degree has become larger than km are moved inside $\mathcal{S}$. The analysis of the evolution of $|\mathcal{S}|$ is based again on the definition of a random process that describes the algorithm dynamics and on the proof that such process behaves in a predictable way for large t.

Let $n = (k-1)m + 1$. For each $i \in \{0, \dots, n-1\}$ and $t > 0$, define $Y_t^i = |V_{m+i} \setminus \mathcal{S}|$ in $G_t^{C,m}$ (V_i is the set of vertices of degree i) before v_t is added to the graph. Let Y_t^n denote the total degree inside $\mathcal{S}$ (i.e. $Y_t^n = \sum_{v \in \mathcal{S}} |\Gamma(v)|$) and X_t the size of the dominating set before v_t is added to the graph. The state of the system, at each step t, is modelled by the (random) vector $(Y_t^0, \dots, Y_t^n, X_t)$. Notice that, for each $t > 0$, the variation in each of the variables is at most m. Also, $Y_t^n + \sum_{i=0}^{(k-1)m}(m+i)Y_t^i = 2mt$, and, at each step $t \geq 1$, when v_t is created the probability that it hits a vertex of degree $m+i$, for $i \in \{0, \dots, (k-1)m\}$ (resp. the dominating set) in any one of the m trials available to it is approximately (omitting $o(1)$ factors, for t large) equal to $P_i = \frac{((m+i-1)(1-\delta_{i,n})+1)Y_t^i}{2mt}$ (where $\delta_{i,n} = 1$ if $i = n$ and zero otherwise).

For $d \in \{0, \dots, n-1\}$ (resp. $d = n$), let E_d denote the event "v_t missed $\mathcal{S}$ and the maximum degree in $\Gamma(v_t)$ is $m + d$" (resp. "v_t did not miss $\mathcal{S}$"). The expected change to Y_t^i, conditioned to the process history up to time t can be computed by further conditioning on the family of events $(E_d)_{d\in\{0,\ldots,n\}}$. We can write such quantity as

$$\sum_{d=0}^{n} E(Y_{t+1}^i - Y_t^i \mid E_d) \Pr[E_d].$$

For t large, the probability in the expression above is approximately $\chi_d = (S_0^d)^m - (S_0^{d-1})^m$ (notation S_a^b stands for $P_a + \ldots + P_b$, with $S_a^b \equiv 0$ if $a > b$). Furthermore, we can approximate $E(Y_{t+1}^i - Y_t^i \mid E_d)$ by the expression:

$$\sum C(h_0, \ldots, h_d, 0, \ldots, 0) \frac{m!}{h_0! \, h_1! \, \ldots \, h_d!} \prod_{i=0}^{d} P_i^{h_i} \frac{1}{\chi_d}$$

where C has $n+1$ arguments, the righmost $n-d$ of which are zero, the sum is over all possible ordered tuples of values $h_0, \ldots, h_d$ such that $\sum_{i=0}^{d} h_i = m$ and $h_d > 0$, and $C(h_0, \ldots, h_d, 0, \ldots, 0)$ contains:

- a term for the addition of v_t to $G_{t-1}^{C,m}$;
- a term $\phi_{i,d}$ for the change to Y_t^i due to the handling of the chosen vertex of maximum degree $m + d$ in $\Gamma(v_t)$ (for $d = n$ this is just one of the vertices hitting $\mathcal{S}$), and
- a term $\psi_{i,s}$ for the change to Y_t^i due to the handling of a vertex accounted for by Y_t^s in $\Gamma(v_t)$, for $s \leq d$.

The first of these is simply $\delta_{i,0}$. We also have

$$\phi_{i,d} = \delta_{d,n} \times \delta_{i,d} + (1 - \delta_{d,n}) \times \{(m + d + 1)\delta_{i,n} - \delta_{i,d}\}$$

(if $d = n$ (i.e. if v_t hits the dominating set) then one is added to Y_t^n, otherwise Y_t^d is decreased and $m + d + 1$ units are added to Y_t^n), and

$$\psi_{i,s} = \delta_{s,n} \times \delta_{i,s} + (1 - \delta_{s,n}) \times \{((m + s)\delta_{s,n-1} + 1)\delta_{i,s+1} - \delta_{i,s}\}$$

(if $s = n$ then Y_t^n is increased by one, if $s = n - 1$ the newly created vertex of degree n must be added to $\mathcal{S}$, and finally, in any other case the vertex that has been hit is moved from V_{m+s} to V_{m+s+1}). Therefore

$$C(h_0, \ldots, h_d, 0, \ldots, 0) = \delta_{i,0} + \phi_{i,d} + (h_d - 1)\psi_{i,d} + \sum_{s=0}^{d-1} h_s \psi_{i,s}.$$

Doing all the sums, for each $i \in \{0, \ldots, n\}$, the expected change to Y_t^i is approximately equal to

$$\delta_{i,0} + \sum_{d=0}^{n} \left\{ \chi_d \phi_{i,d} + \psi_{i,d} \left[mP_d(S_0^d)^{m-1} - \chi_d \right] + \frac{m[\chi_d - P_d(S_0^d)^{m-1}]}{S_0^{d-1}} \sum_{s=0}^{d-1} \psi_{i,s} P_s \right\}.$$

Also $E(X_{t+1} - X_t)$ is approximately equal to

$$\frac{km E(Y_t^{n-1})}{2t} + E\left[\left(1 - \frac{Y_t^n}{2mt} - \frac{k Y_t^{n-1}}{2t} \right)^m \right].$$

The value α_{up}^C reported in Table 1 is defined by $\alpha = kmy^{n-1} + (1 - \frac{y^n}{2m} - \frac{ky^{n-1}}{2})^m$ where y^i, for $i \in \{0, \ldots, n\}$ satisfy $y^i = \left[\mathrm{E}(Y_{t+1}^i) - \mathrm{E}(Y_t^i)\right]\big|_{Y^i = y^i t, i \in \{0,\ldots,n\}}$. The parameter k can be chosen arbitrarily (larger values of k give slightly better values of α_{up}^C). For each $m \le 7$ values of $k \le 20$ give the results in Table 1.

6 Generalisations

The algorithms presented in the previous sections generalise naturally to larger values of h. We present here the generalisation of Algorithm 2 for finding an h-dominating set in the random graph process and of Algorithm 3 for the pure copy process. We then briefly sketch the analysis of the first heuristic, and comment on the numerical results presented in Table 3.

Algorithm 4. A vertex in the h-dominating set can be either *permanent* (set $\mathcal{P}$) as it will never be removed from the set or *i-removable* (set $\mathcal{R}_i$) meaning that it is dominated $i \in \{0, \ldots, h-1\}$ times by other permanent vertices.

Before the first step of the algorithm the graph consists of a single vertex v_0 and $\mathcal{R}_0 = \{v_0\}$. After v_t is created and connected to m neighbours, v_t is added to $\mathcal{R}_0$ if $\Gamma(v_t) \cap (\mathcal{P} \cup \bigcup_i \mathcal{R}_i) = \emptyset$, otherwise if $\Gamma(v_t) \cap \mathcal{P} = \emptyset$, v_t is added to $\mathcal{P}$, any vertex in $\Gamma(v_t) \cap \mathcal{R}_i$, for $i < h-1$ is moved to $\mathcal{R}_{i+1}$ and any vertex in $\Gamma(v_t) \cap \mathcal{R}_{h-1}$ is moved to $V \setminus \mathcal{S}$.

The process in Algorithm 4 can be modelled by h sequences of random variables: R_t^i for $i \in \{0, \ldots, h-1\}$ with $R_t^i = |\mathcal{R}_i|$ in the graph $G_{t-1}^{R,m}$ just before v_t is added to it, and $P_t = |\mathcal{P}|$. The expected changes in these variables satisfy

$$\mathrm{E}(P_{t+1}) = \mathrm{E}(P_t) + \mathrm{E}[(1 - \tfrac{P_t}{t})^m] - \mathrm{E}[(1 - \tfrac{P_t}{t} - \textstyle\sum_i \tfrac{R_t^i}{t})^m]$$
$$\mathrm{E}(R_{t+1}^i) = \mathrm{E}(R_t^i) + \mathrm{E}[(1 - \tfrac{P_t}{t} - \textstyle\sum_i \tfrac{R_t^i}{t})^m]\delta_{i,0} + m\mathrm{E}[(\tfrac{R_t^{i-1}}{t}(1 - \delta_{i,0}) - \tfrac{R_t^i}{t})(1 - \tfrac{P_t}{t})^{m-1}]$$

therefore, we have $P_t \sim pt$ where p satisfies:

$$p = (1-p)^m - \{1 - [(1-p)^m - p][1 - (\frac{m(1-p)^{m-1}}{1 + m(1-p)^{m-1}})^h] - p\}^m$$

and $R_t^i \sim r^i t$ where $r^0 = \frac{(1-p)^m - p}{1 + m(1-p)^{m-1}}$, and $r^i = \frac{m(1-p)^{m-1}}{1 + m(1-p)^{m-1}}r^{i-1}$. The values reported in Table 3 below are given by $p + \sum_{i=0}^{h-1} r^i$.

Table 3. Upper bounds on γ_h/t, for $h > 1$

m	$h = 2$	$h = 3$	$h = 4$	$h = 5$	m	$h = 2$	$h = 3$	$h = 4$	$h = 5$
2	0.4484				2	0.5			
3	0.368	0.3836			3	0.4075	0.6		
4	0.3162	0.3307	0.332		4	0.3359	0.4852	0.6663	
5	0.2795	0.2929	0.2944	0.2931	5	0.282	0.4073	0.5423	0.7036
6	0.2517	0.2641	0.2658	0.2647	6	0.2428	0.3523	0.4613	0.5862
7	0.2298	0.2413	0.2431	0.2422	7	0.2132	0.3066	0.4035	0.5056

Algorithm 4.   Algorithm 5.

Algorithm 5. Before the first step of the algorithm the graph consists of a single vertex v_0 and $S = \{v_0\}$. After v_t is created and connected to m neighbours, the set Z of all newly generated vertices of degree more than km are added to S. If v_t is dominated $h - x$ times by elements of S then the x vertices of highest degree in $\Gamma(v_t) \setminus Z$ are added to the dominating set.

7 Tightness of the Algorithmic Results

An interesting and simple argument can be used to complement the algorithmic results presented in the previous sections. The argument is based on the following result.

Lemma 4. *Let S be an h-dominating set in a graph $G = (V, E)$. If the total degree of the vertices in S is at least d, then $|S| \geq \sum_{i > i_0} |V_i|$, where i_0 is the largest index i for which $\sum_{j > i} j|V_j| \geq d$.*

Proof. Let i_0 be defined as in the statement of the result. If $\Delta = \max_{v \in V} |\Gamma(v)|$, then the set $V_{i_0} \cup \ldots \cup V_\Delta$ is the smallest set of vertices in G with total degree at least d (any other vertex in G would have smaller degree and therefore it would contribute less to the total degree).   □

The total degree of an h-dominating set in $G_t^{M,m}$ must be at least $h(t - |S|) \geq ht(1 - \alpha_{up}^M)$. Hence a lower bound on the size of any dominating set in a web-graph is obtained by using information on the proportional degree sequence.

In the pure copy web graphs Bollobas *et al.* [5] (see also Cooper [8]) proved that $|V_i| = tn_i + O(\sqrt{t \log t})$ a.a.s. for any $i \geq m$, where

$$n_i = \frac{2m(m+1)}{i(i+1)(i+2)}.$$

In the same paper (p. 288) it is possible to find a result about $G_t^{R,m}$. In the random graph process, for any $i \geq m$,

$$n_i = \frac{1}{m+1} \left(\frac{m}{m+1} \right)^{i-m}.$$

Again it is possible to prove that $|V_i|$ is concentrated around $n_i t$. For $m > 1$, the bounds reported in Table 1 in Section 2 are obtained using Lemma 4 and the approximations above for $|V_i|$ in each case. Bounds for $h > 1$ are in Table 4.

8 Trees

For $m = 1$ the graph processes under consideration generate a connected graph without cycles. Such structural property can be exploited to obtain improved lower bounds on γ_1. Without loss of generality, any vertex in such graphs that has at least one neighbour $u \in V_1$ must be part of a minimum size dominating set. The number of such vertices is precisely $t - |V_1| - |I|$ where $|I|$ is the number of

Table 4. Lower bounds on γ_h/t, for $h > 1$

m	$h = 2$	$h = 3$	$h = 4$	$h = 5$	m	$h = 2$	$h = 3$	$h = 4$	$h = 5$
2	0.1317				2	0.0545			
3	0.1001	0.178			3	0.0316	0.0351		
4	0.0687	0.1342	0.1678		4	0.023	0.0333	0.0246	
5	0.0649	0.0935	0.1346	0.1938	5	0.0183	0.0284	0.0302	0.0183
6	0.0535	0.0849	0.1155	0.1573	6	0.0141	0.0233	0.0299	0.0269
7	0.0406	0.0692	0.1033	0.1349	7	0.0113	0.0203	0.0271	0.0283

Random graph.   Pure copy.

Table 5. Average values obtained over 1000 experiments for each value of t

t	$\gamma_1(G_{1,t}^R)/t$	$\gamma_1(G_{1,t}^C)/t$
10000	3745.053	2943.157
20000	7489.3	5887.301
30000	11233.68	8829.288
40000	14980.384	11772.175
50000	18725.448	14714.073
...		
100000	37451.20312	29424.216

vertices that have no neighbour in V_1. The cardinality of I can be estimated in both models via either a martingale argument similar to those used in previous sections or through the technique exploited in [8] to estimate $|V_i|$. The lower bounds in Table 1 for $m = 1$ come from this argument. Details of such analysis are left for the final version of this paper.

We end this Section reporting on some simple empirical results which help put the mathematical analysis performed so far into context. It is well known [7] that minimum size dominating sets can be found efficiently in trees. We implemented Cockayne et al's algorithm and tested its performance. For different values of t, we repeatedly ran the two graph processes up to time t, and then applied the optimal algorithm mentioned above. Table 5 reports the average values we obtained. The least square approximation lines over the full set of data we collected are (coefficients rounded to the sixth decimal place) $y = 0.374509x - 0.214185$ for the random graph case, and $y = 0.294294x + 0.32284$ for the pure copy case. These results indicate that our algorithms are able to get better results for graphs generated according to the pure copy process ($\alpha_{up}^C = 0.3333$) than for graphs generated by the other process ($\alpha_{up}^R = 0.5$). We leave the finding of improved algorithms especially in the random graph process or indeed better lower bounds in either models as an interesting open problem of this work.

References

1. N. Alon, J. H. Spencer, and P. Erdős. *The Probabilistic Method*. John Wiley & Sons, 1992.

2. K. Alzoubi, P. J. Wan, and O. Frieder. Message-optimal connected dominating sets in mobile ad hoc networks. In *Proc. 3rd ACM Internat. Symp. on Mobile Ad-hoc Networking & Computing*, pp 157–164, 2002.
3. A. Barabási and R. Albert. Emergence of scaling in random networks. *Science*, 286:509–512, 1999.
4. E. A. Bender and E. R. Canfield. The asymptotic number of labeled graphs with given degree sequences. *JCT*, A 24:296–307, 1978.
5. B. Bollobás, O. Riordan, J. Spencer, and G. Tusnády. The degree sequence of a scale-free random graph process. *RSA*, 18:279–290, 2001.
6. A. Broder, R. Kumar, F. Maghoul, P. Raghavan, S. Rajagopalan, R. Stata, A. Tomkins, and J. Wiener. Graph structure in the web. In *Proc. 9th WWW*, pp 309–320, 2000.
7. E. Cockayne, S. Goodman, and S. Hedetniemi. A linear algorithm for the domination number of a tree. *IPL*, 4:41–44, 1975.
8. C. Cooper. The age specific degree distribution of web-graphs. Submitted to Combinatorics Probability and Computing, 2002.
9. C. Cooper and A. Frieze. A general model of web graphs. *RSA*, 22:311–335, 2003.
10. W. Duckworth and M. Zito. Sparse hypercube 3-spanners. *DAM*, 103:289–295, 2000.
11. S. Eidenbenz. Online dominating set and variations on restricted graph classes. Technical Report 380, Department of Computer Science, ETH Zürich, 2002.
12. U. Feige. A threshold of ln n for approximating set cover. *JACM*, 45:634–652, 1998.
13. M. R. Garey and D. S. Johnson. Strong NP-Completeness results: Motivation, examples, and implications. *Journal of the Association for Computing Machinery*, 25(3):499–508, 1978.
14. F. Harary and T. W. Haynes. Double domination in graphs. *Ars Comb.*, 55:201–213, 2000.
15. T. W. Haynes, S. T. Hedetniemi, and P. J. Slater, editors. *Domination in Graphs: Advanced Topics*. Marcel Dekker, 1998.
16. T. W. Haynes, S. T. Hedetniemi, and P. J. Slater. *Fundamentals of Domination in Graphs*. Marcel Dekker, 1998.
17. G.-H. King and W.-G. Tzeng. On-line algorithms for the dominating set problem. *IPL*, 61:11–14, 1997.
18. R. Klasing and C. Laforest. Hardness results and approximation algorithms of k-tuple domination in graphs. *IPL*, 89:75–83, 2004.
19. R. Kumar, P. Raghavan, S. Rajagopalan, D. Sivakumar, A. Tomkins, and E. Upfal. The web as a graph. In *Proc. ACM Symp. on PODS*, pp 1–10, 2000.
20. M. Levene and R. Wheeldon. Web dynamics. *Software Focus*, 2:31–38, 2001.
21. C.-S. Liao and G. J. Chang. k-tuple domination in graphs. *IPL*, 87:45–50, 2003.
22. I. Stojmenovic, M. Seddigh, and J. Zunic. Dominating sets and neighbor elimination-based broadcasting algorithms in wireless networks. *IEEE Trans. Parallel and Dist. Systems*, 13:14–25, 2002.
23. D. J. Watts and S. H. Strogatz. Collective dynamics of 'small-world' networks. *Nature*, 393(4):440–442, June 1998.
24. B. Wieland and A. P. Godbole. On the domination number of a random graph. *Elec. J. Combinat.*, 8:# R37, 2001.
25. N. C. Wormald. The differential equation method for random graph processes and greedy algorithms. In M. Karoński and H. J. Prömel, editors, *Lectures on Approximation and Randomized Algorithms*, pages 73–155. PWN, Warsaw, 1999.
26. M. Zito. Greedy algorithms for minimisation problems in random regular graphs. *Proc 9th ESA*, pp 524–536, LNCS 2161. Springer-Verlag, 2001.

A Geometric Preferential Attachment Model of Networks

Abraham D. Flaxman, Alan M. Frieze*, and Juan Vera

Department of Mathematical Sciences,
Carnegie Mellon University,
Pittsburgh PA15213,
U.S.A.

Abstract. We study a random graph G_n that combines certain aspects of geometric random graphs and preferential attachment graphs. The vertices of G_n are n sequentially generated points $x_1, x_2, \ldots, x_n$ chosen uniformly at random from the unit sphere in $\mathbf{R}^3$. After generating x_t, we randomly connect it to m points from those points in $x_1, x_2, \ldots, x_{t-1}$ which are within distance r. Neighbours are chosen with probability proportional to their current degree. We show that if m is sufficiently large and if $r \geq \log n / n^{1/2-\beta}$ for some constant β then **whp** at time n the number of vertices of degree k follows a power law with exponent 3. Unlike the preferential attachment graph, this geometric preferential attachment graph has small separators, similar to experimental observations of [7]. We further show that if $m \geq K \log n$, K sufficiently large, then G_n is connected and has diameter $O(m/r)$ **whp**.

1 Introduction

Recently there has been much interest in understanding the properties of real-world large-scale networks such as the structure of the Internet and the World Wide Web. For a general introduction to this topic, see Bollobás and Riordan [8], Hayes [21], Watts [32], or Aiello, Chung and Lu [2]. One approach is to model these networks by random graphs. Experimental studies by Albert, Barabási, and Jeong [3], Broder et al [12], and Faloutsos, Faloutsos, and Faloutsos [20] have demonstrated that in the World Wide Web/Internet the proportion of vertices of a given degree follows an approximate inverse power law i.e. the proportion of vertices of degree k is approximately $Ck^{-\alpha}$ for some constants C, α. The classical models of random graphs introduced by Erdős and Renyi [18] do not have power law degree sequences, so they are not suitable for modeling these networks. This has driven the development of various alternative models for random graphs.

One approach is to generate graphs with a prescribed degree sequence (or prescribed expected degree sequence). This is proposed as a model for the web graph by Aiello, Chung, and Lu in [1]. Mihail and Papadimitriou also use this model [27] in their study of large eigenvalues, as do Chung, Lu, and Vu in [14].

---

* Supported in part by NSF grant CCR-0200945.

S. Leonardi (Ed.): WAW 2004, LNCS 3243, pp. 44–55, 2004.
© Springer-Verlag Berlin Heidelberg 2004

An alternative approach, which we will follow in this paper, is to sample graphs via some generative procedure which yields a power law distribution. There is a long history of such models, outlined in the survey by Mitzenmacher [29]. We will use an extension of the preferential attachment model to generate our random graph. The preferential attachment model has been the subject of recently revived interest. It dates back to Yule [33] and Simon [31]. It was proposed as a random graph model for the web by Barabási and Albert [4], and their description was elaborated by Bollobás and Riordan [9] who showed that at time n, **whp** the diameter of a graph constructed in this way is asymptotic to $\frac{\log \log n}{\log n}$. Subsequently, Bollobás, Riordan, Spencer and Tusnády [11] proved that the degree sequence of such graphs does follow a power law distribution.

The random graph defined in the previous paragraph has good expansion properties. For example, Mihail, Papadimitriou and Saberi [28] showed that **whp** the preferential attachment model has conductance bounded below by a constant. This is in contrast to what has sometimes been found experimentally, for example by Blandford, Blelloch and Kash [7]. Their results seem to suggest the existence of smaller separators than implied by random graphs with the same average degree. The aim of this paper is to describe a random graph model which has *both* a power-law degree distribution and which has small separators.

We study here the following process which generates a sequence of graphs $G_t, t = 1, 2, \ldots, n$. The graph $G_t = (V_t, E_t)$ has t vertices and e_t edges. Here V_t is a subset of S, the surface of the sphere in $\mathbf{R}^3$ of radius $\frac{1}{2\sqrt{\pi}}$ (so that $area(S) = 1$).

For $u \in S$ and $r > 0$ we let $B_r(u)$ denote the spherical cap of radius r around u in S. More precisely, $B_r(u) = \{x \in S : ||x - u|| \leq r\}$.

1.1   The Random Process

- **Time Step 1:**   To initialize the process, we start with G_1 containing a single vertex x_1 chosen at random in S. The edge (multi)set consists of m loops at x_1.
- **Time Step $t+1$:**   We choose vertex x_{t+1} uniformly at random in S and add it to G_t. If $V_t \cap B_r(x_{t+1})$ is nonempty, we add m random edges (x_{t+1}, y_i), $i = 1, 2, \ldots, m$ incident with x_{t+1}. Here, each y_i is chosen from $V_t \cap B_r(x_{t+1})$ and for $x \in V_t \cap B_r(x_{t+1})$,

$$\mathbf{Pr}(y_i = x) = \frac{\deg_t(x)}{D_t(B_r(x_{t+1}))},$$

where $\deg_t(x)$ denotes the degree of vertex x in G_t and $V_t(U) = V_t \cap U$ and $D_t(U) = \sum_{x \in V_t(U)} \deg_t(x)$.

If $V_t \cap B_r(x_{t+1})$ is empty then we add m loops at x_{t+1}.

Let $d_k(t)$ denote the number of vertices of degree k at time t.

We will prove the following:

Theorem 1.

(a) *If $0 < \beta < 1/2$ is constant and $r \geq n^{\beta-1/2} \log n$ and m is a sufficiently large constant then there exists a constant $c > 0$ such that* **whp**

$$d_k(n) = \frac{cn}{k(k+1)(k+2)} + O(n^{1-\gamma}), [1]$$

for some $0 < \gamma < 1$.

(b) *If $r = o(1)$ then* **whp** *V_n can be partitioned into $T, \bar{T}$ such that $|T|, |\bar{T}| \sim n/2$, and there are at most $4\sqrt{\pi} rnm$ edges between T and $\bar{T}$.*

(c) *If $r \geq n^{-1/2} \log n$ and $m \geq K \log n$ and K is sufficiently large then* **whp** *G_n is connected.*

(d) *If $r \geq n^{-1/2} \log n$ and $m \geq K \log n$ and K is sufficiently large then* **whp** *G_n has diameter $O(\log n/r)$.*

We note that geometric models of trees with power laws have been considered in [19], [5] and [6].

1.2   Some Definitions

There exists some constant c_0 such that for any $u \in S$, we have

$$A_r = Area(B_r(u)) = c_0 n^{2\beta-1} (\log n)^2.$$

Given $u \in S$, we define

$$V_t(u) = V_t(B_r(u)) \text{ and } D_t(u) = D_t(B_r(u)).$$

Given $v \in V_t$, we have $\deg_t(v) = m + \deg_t^-(v)$, where $\deg_t^-(v)$ is the number of edges of G_t that are incident to v and were added by vertices that chose v as a neighbor.

Given $U \subseteq S$, we define $D_t^-(U) = \sum_{v \in V_t(U)} \deg_t^-(v)$. We also define $D_t^-(u) = D_t^-(B_r(u))$. Notice that $D_t(U) = m|V_t(U)| + D_t^-(U)$.

We write $\bar{d}_k(t)$ to denote the expectation of $d_k(t)$. We also localize these notions: given $U \subseteq S$ and $u \in S$ we define $d_k(t, U)$ to be the number of vertices of degree k at time t in U and $d_k(t, u) = d_k(t, B_r(u))$.

2   Small Separators

Theorem 1 part (b) is the easiest part to prove. We use the geometry of the instance to obtain a sparse cut. Consider partitioning the vertices using a great circle of S. This will divide V into sets T and $\bar{T}$ which each contain about $n/2$ vertices. More precisely, we have

$$\mathbf{Pr}\left[|T| < (1-\epsilon)n/2\right] = \mathbf{Pr}\left[|\bar{T}| < (1-\epsilon)n/2\right] \leq e^{-\epsilon^2 n/6}.$$

---

[1] Asymptotics are taken as $n \to \infty$.

Since edges only appear between vertices within distance r, only vertices appearing in the strip within distance r of the great circle can appear in the cut. Since $r = o(1)$, this strip has area less than $3r\sqrt{\pi}$, so, letting U denote the vertices appearing in this strip, we have

$$\mathbf{Pr}\left[|U| \geq 4\sqrt{\pi}rn\right] \leq e^{-\sqrt{\pi}rn/9}$$

vertices. Even if every one of the vertices chooses its m neighbors on the opposite side of the cut, this will yield at most $4\sqrt{\pi}rnm$ edges **whp**. So the graph has a cut with $\frac{e(T,\bar{T})}{|T||\bar{T}|} \leq \frac{17\sqrt{\pi}rm}{n}$ with probability at least $1 - e^{-\Omega(rn)}$.

3   Proving a Power Law

3.1   Establishing a Recurrence for $\bar{d}_k(t)$

Our approach to proving Theorem 1 part (a) is to find a recurrence for $\bar{d}_k(t)$.

We define $\bar{d}_{m-1}(t) = 0$ for all integers t with $t > 0$. Let $\eta_1(t)$ denote the probability that $V_t \cap B_r(x_{t+1}) = \emptyset$ so $\eta_1(t) = (1 - A_r)^t$. Let $\eta_2(t)$ denote the probability that a parallel edge is created. Thus

$$\eta_2(t) = O\left(\sum_{i=m}^{k} d_i(t, x_{t+1}) i^2 / D_t(x_{t+1})^2\right) = O(k/D_t(x_{t+1})).$$

Then for $k \geq m$,

$$\mathbf{E}\left[d_k(t+1) \mid G_t, x_{t+1}\right] = d_k(t) + m d_{k-1}(t, x_{t+1}) \frac{k-1}{D_t(x_{t+1})}$$

$$- m d_k(t, x_{t+1}) \frac{k}{D_t(x_{t+1})} + 1_{k=m} + O(\eta_1(t) + \eta_2(t)). \quad (1)$$

Let $\alpha = \frac{1}{400}$ and $\gamma = \frac{\alpha^2}{2(1-\alpha^2)}$, and let $\mathcal{A}_t$ be the event $\{|D_t(x_{t+1}) - 2mA_r t| < A_r m t^{1-\gamma}\}$.

Then, because $\mathbf{E}[d_k(t, x_{t+1})] \leq k^{-1}\mathbf{E}[m|V_t(B_{2r}(x_{t+1}))|] \leq k^{-1}m(4A_r t)$ and $d_k(t, x_{t+1}) \leq k^{-1}D_t(x_{t+1})$, we have

$$\mathbf{E}\left[\frac{d_k(t, x_{t+1})}{D_t(x_{t+1})}\right]$$

$$= \mathbf{E}\left[\frac{d_k(t, x_{t+1})}{D_t(x_{t+1})} \mid \mathcal{A}_t\right]\mathbf{Pr}\left[\mathcal{A}_t\right] + \mathbf{E}\left[\frac{d_k(t, x_{t+1})}{D_t(x_{t+1})} \mid \neg\mathcal{A}_t\right]\mathbf{Pr}\left[\neg\mathcal{A}_t\right]$$

$$= \frac{\mathbf{E}\left[d_k(t, x_{t+1}) \mid \mathcal{A}_t\right]}{2mA_r t}\mathbf{Pr}\left[\mathcal{A}_t\right] + O\left(\frac{t^{-\gamma}}{k}\right) + O\left(\frac{\mathbf{Pr}\left[\neg\mathcal{A}_t\right]}{k}\right)$$

$$= \frac{\mathbf{E}\left[d_k(t, x_{t+1})\right]}{2mA_r t} + O\left(\frac{t^{-\gamma}}{k}\right)$$

$$+ \left(O\left(\frac{1}{k}\right) - \frac{\mathbf{E}\left[d_k(t, x_{t+1}) \mid \neg\mathcal{A}_t\right]}{2mA_r t}\right)\mathbf{Pr}\left[\neg\mathcal{A}_t\right]$$

$$= \frac{\mathbf{E}\left[d_k(t, x_{t+1})\right]}{2mA_r t} + O\left(\frac{t^{-\gamma}}{k}\right) + O\left(\frac{1}{k} + \frac{1}{A_r}\right)\mathbf{Pr}\left[\neg\mathcal{A}_t\right].$$

In Lemmas 1 and 3 below we prove that $\mathbf{E}\left[d_k(t, x_{t+1})\right] = A_r \bar{d}_k(t)$ and that if $t \geq n^{1-\alpha}$ then $\mathbf{Pr}\left[\neg \mathcal{A}_t\right] = O\left(n^{-2}\right)$. Therefore if $t \geq n^{1-\alpha}$ then

$$\mathbf{E}\left[\frac{d_k(t, x_{t+1})}{D_t(x_{t+1})}\right] = \frac{\bar{d}_k(t)}{2mt} + O\left(\frac{t^{-\gamma}}{k}\right). \tag{2}$$

In a similar way

$$\mathbf{E}\left[\frac{d_{k-1}(t, x_{t+1})}{D_t(x_{t+1})}\right] = \frac{\bar{d}_{k-1}(t)}{2mt} + O\left(\frac{t^{-\gamma}}{k-1}\right). \tag{3}$$

Now note that

$$\eta_2(t) \leq \mathbf{Pr}(\neg \mathcal{A}_t) + O(k/tA_r).$$

Taking expectations on both sides of Eq. (1) and using Eq. (2) and Eq. (3), we see that if $k \leq n^{1-\alpha} \leq t$ then

$$\bar{d}_k(t+1) = \bar{d}_k(t) + \frac{k-1}{2t}\bar{d}_{k-1}(t) - \frac{k}{2t}\bar{d}_k(t) + 1_{k=m} + O\left(t^{-\gamma}\right) \tag{4}$$

We consider the recurrence given by $f_{m-1} = 0$ and for $k \geq m$,

$$f_k = 1_{k=m} + \frac{k-1}{2}f_{k-1} - \frac{k}{2}f_k$$

which has solution

$$f_k = f_m \prod_{i=m+1}^{k} \frac{i-1}{i+2}$$
$$= f_m \frac{m(m+1)(m+2)}{k(k+1)(k+2)}.$$

Let $t_0 = n^{1-\alpha}$. We finish the proof of Theorem 1(a) by showing that there exists a constant $M > 0$ such that

$$|\bar{d}_k(t) - f_k t| \leq M(t_0 + t^{1-\gamma}) \tag{5}$$

for all k with $m \leq k \leq n^{1-\alpha}$ and all $t > 0$. For $k > n^{1-\alpha}$ we use the fact that $\bar{d}_k(t) \leq 2mt/k$.

Let $\Theta_k(t) = \bar{d}_k(t) - f_k t$. Then for $m \leq k \leq n^\alpha$ and $t \geq t_0$,

$$\Theta_k(t+1) = \frac{k-1}{2t}\Theta_{k-1}(t) - \frac{k}{2t}\Theta_k(t) + O(t^{-\gamma}). \tag{6}$$

Let L denote the hidden constant in $O(t^{-\gamma})$ of (6). Our inductive hypothesis $\mathcal{H}_t$ is that $|\Theta_k(t)| \leq M(t_0 + t^{1-\gamma})$ for every $m \leq k \leq n^{1-\alpha}$. It is trivially true for $t \leq t_0$. So assume that $t \geq t_0$. Then, from (6),

$$|\Theta_k(t+1)| \leq M(t_0 + t^{1-\gamma}) + Lt^{-\gamma}$$
$$\leq M(t_0 + (t+1)^{1-\gamma}).$$

This verifies $\mathcal{H}_{t+1}$ and completes the proof by induction.

3.2 Expected Value of $d_k(t, u)$

Lemma 1. *Let $u \in S$ and let k and t be positive integers. Then $\mathbf{E}\left[d_k(t, u)\right] = A_r \overline{d}_k(t)$.*

Proof. Left to the full version:
see http://www.math.cmu.edu/ aflp/html/GeoWeb.ps.   □

Lemma 2. *Let $u \in S$ and $t > 0$ then $\mathbf{E}\left[D_t(u)\right] = 2A_r mt$.*

Proof.

$$\mathbf{E}\left[D_t(u)\right] = \sum_{k>0} \mathbf{E}\left[d_k(t, u)\right] = A_r \sum_{k>0} \mathbf{E}\left[d_k(t)\right] = A_r \mathbf{E}\left[\sum_{k>0} d_k(t)\right] = 2A_r mt.$$

□

3.3 Concentration of $D_t(u)$

In this section we prove

Lemma 3. *Let $\alpha = 1/400$ and $\gamma = \frac{\alpha^2}{2(1-\alpha^2)}$ and $n_0 = n^{1-2\alpha}$. If $t > n^{1-\alpha}$ and $u \in S$ then*

$$\mathbf{Pr}\left[|D_t(u) - \mathbf{E}\left[D_t(u)\right]| \geq A_r mt^{1-\gamma}\right] = O\left(n^{-2}\right).$$

Proof. We think of every edge added as two directed edges. Then choosing a vertex by preferential attachment is equivalent to choosing one of these directed edges uniformly, and taking the vertex pointed to by this edge as the chosen vertex. So the ith step of the process is defined by a tuple of random variables $T = (X, Y_1, \ldots, Y_m) \in S \times E_i^m$ where X is the location of the new vertex, a randomly chosen point in S, and Y_j is an edge chosen u.a.r. among the edges directed into $B_r(X)$ in G_{i-1}. The process G_t is then defined by a sequence $\langle T_1, \ldots, T_t \rangle$, where each $T_i \in S \times E_i^m$.

Let s be a sequence $s = \langle s_1, \ldots, s_t \rangle$ where $s_i = (x_i, y_{(i-1)m+1}, \ldots, y_{im})$ with $x_i \in S$ and $y_j \in E_{\lceil t/j \rceil}$. We say s is *acceptable* if for every j, y_j is an edge entering $B_r(x_{\lceil t/j \rceil})$. Notice that non-acceptable sequences have probability 0 of being observed.

In what follows we condition on the event

$$\mathcal{E} = \{\text{for all } s \text{ with } n_0 < s \leq n \text{ we have } D_s(u) > (1+\alpha)A_r ms\},$$

where $\alpha > 0$ is an appropriate constant that will be chosen later.

Fix $t > 0$. Fix an acceptable sequence $s = \langle s_1, \ldots, s_t \rangle$, and let $A_k(s) = \{z \in S \times E_k^m : \langle s_1, \ldots, s_{k-1}, z \rangle \text{ is acceptable}\}$. For any k with $1 \leq k \leq t$ and any $z \in A_k(s)$ let

$$g_k(z) = \mathbf{E}\left[D_t(u) \mid T_1 = s_1, \ldots, T_{k-1} = s_{k-1}, T_k = z, \mathcal{E}\right],$$

let $r_k(s) = \sup\{|g_k(z_1) - g_k(z_2)| : z_1, z_2 \in A_k(s)\}$ and let $R^2(s) = \sum_{k=1}^{t} r_k(s)^2$. Finally define $\hat{r}^2 = \sup_s R^2(s)$, where the supremum is taken over all acceptable sequences.

By Thm 3.7 of [26] we know that for all $\lambda > 0$,

$$\mathbf{Pr}\left[|D_t(u) - \mathbf{E}\left[D_t(u)\right]| \geq \lambda\right] < 2e^{-2\lambda^2/\hat{r}^2} + \mathbf{Pr}\left[\neg\mathcal{E}\right]. \tag{7}$$

Now, fix k, with $1 \leq k \leq t$. Our goal now is to bound $r_k(s)$ for any acceptable sequence s.

Fix $z, z' \in A_k(s)$. We define $\Omega(G_t, G_t')$, the following coupling between $G_t = G_t(s_1, \ldots, s_{k-1}, z)$ and $G_t' = G_t(s_1, \ldots, s_{k-1}, z')$

- Step k: Start with the graph $G_k(s_1, \ldots, s_{k-1}, z)$ and $G_k'(s_1, \ldots, s_{k-1}, z')$ respectively.
- Step τ ($\tau > k$): Choose the same point $x_\tau \in S$ in both processes. Let E_τ (resp. E_τ') be the edges pointing to the vertices in $B_r(x_\tau)$ in $G_{\tau-1}$ (resp. $G_{\tau-1}'$). Let $C_\tau = E_\tau \cap E_\tau'$, $R_\tau = E_\tau \setminus E_\tau'$, and $L_\tau = E_\tau' \setminus E_\tau$
  Let $D_\tau = |E_\tau|$ and $D_\tau' = |E_\tau'|$. Without loss of generality assume that $D_\tau \leq D_\tau'$. Note that $D_\tau = 0$ iff $V_{\tau-1} \cap B_r(x_\tau) = \emptyset$, in which case $D_\tau' = 0$ as well. Assume for now that $D_\tau > 0$. Let $p = 1/D_\tau$ and let $p' = 1/D_\tau'$. Construct G_τ choosing m edges u.a.r. $e_1^\tau, \ldots, e_m^\tau$ in E_τ, and joining x_τ to the end point of them. For each of the m edges $e_i = e_i^\tau$, we define $\hat{e}_i = \hat{e}_i^\tau$ by
  - If $e_i \in C_\tau$ then, with probability p'/p, $\hat{e}_i = e_i$. With probability $1 - p'/p$, $\hat{e}_i$ is chosen from L_τ u.a.r.
  - If $e_i \in R_\tau$, $\hat{e}_i \in L_\tau$ is chosen u.a.r.
  Notice that for every $i = 1, \ldots, m$ and every $e \in E_\tau'$, $\mathbf{Pr}\left[\hat{e}_i = e\right] = p'$. To finish, in G_τ' join x_τ to the m vertices pointed to by the edges $\hat{e}_i$.

Lemma 4. *Let* $\Delta_\tau = \Delta_\tau(k, s, z, z', u) = |E_{G_\tau}(B_r(u)) \triangle E_{G_\tau'}(B_r(u))|$, *the discrepancy between the edge-sets incident to* $V_\tau(u)$ *in the two coupled graphs. Then* $|g_k(z) - g_k(z')| \leq \mathbf{E}\left[\Delta_t 1_\mathcal{E}\right]/\mathbf{Pr}\left[\mathcal{E}\right]$.

Proof.

$$\begin{aligned}
|g_k(z) - g_k(z')| &= |\mathbf{E}_{G_t}[D_t(u) \mid \mathcal{E}] - \mathbf{E}_{G_t'}[D_t(u) \mid \mathcal{E}]| \\
&\leq \mathbf{E}_{\Omega(G_t, G_t')}[D_t'(u) - D_t(u) \mid \mathcal{E}] \\
&\leq \mathbf{E}_{\Omega(G_t, G_t')}[\Delta_t \mid \mathcal{E}] \\
&= \mathbf{E}_{\Omega(G_t, G_t')}[\Delta_t 1_\mathcal{E}]/\mathbf{Pr}\left[\mathcal{E}\right].
\end{aligned}$$

$\square$

Lemma 5. *Let* $k_0 = \max\{k, n_0\}$ *and* ϵ_1 *be a small positive constant, then*

$$\mathbf{E}\left[\Delta_t 1_\mathcal{E}\right] \leq 16 m A_r n^{c_0 \epsilon_1} \left(\frac{k_0}{k}\right)^{1/(1-2\epsilon_1)} \left(\frac{t}{k_0}\right)^{1/(1+\alpha)}.$$

Proof. Left to the full version.   □

By applying Lemma 5, we have that for any acceptable sequence we obtain

$$R^2(s) = O\left(A_r^2 m^2 t^{\frac{2}{1+\alpha}} n^{2c_0\epsilon_1} n_0^{\frac{2}{1-2\epsilon_1} - \frac{2}{1+\alpha}}\right),$$

where we rely on the fact that $\mathbf{Pr}\,[\mathcal{E}] = 1 - o(1)$, which is proved below.
Therefore, by using Eq. (7), we have

$$\mathbf{Pr}\left[|D_t(u) - \mathbf{E}\,[D_t(u)]| \geq A_r m t^{\frac{1}{1+\alpha}} n^{c_0\epsilon_1/2} n_0^{\frac{1}{1-2\epsilon_1} - \frac{1}{1+\alpha}} \log n\right]$$
$$\leq e^{-\Omega(\log n^2)} + \mathbf{Pr}\,[\neg\mathcal{E}]. \quad (8)$$

Now we concentrate in bounding $\mathbf{Pr}\,[\neg\mathcal{E}]$.

Lemma 6. *Let* $\alpha = 1/400$. *There is* $c > 0$ *such that* $\mathbf{Pr}\,[\neg\mathcal{E}] = O\left(e^{-cn_0 A_r}\right)$.

Proof. Left to the full version.   □

Returning to (8) and taking ϵ_1 sufficiently small, we see that there is $c > 0$ such that

$$\mathbf{Pr}\left[|D_t(u) - \mathbf{E}\,[D_t(u)]| \geq A_r m t^{1-\gamma}\right] \leq n^{-2} + O\left(e^{-cn_0 A_r}\right), \quad (9)$$

which completes the proof of Lemma 3.

4   Connectivity

Here we are going to prove that for $r \geq n^{-1/2} \log n$, $m > K \log n$, and K sufficiently large, **whp** G_n is connected and has diameter $O(\log n/r)$. Notice that G_n is a subgraph of the graph $G(n, r)$, the intersection graph of the caps $B_r(x_t)$, $t = 1, 2, \ldots, n$ and therefore it is disconnected for $r = o(n^{-1/2} \log n)$ [30]. We denote the diameter of G by $\mathrm{diam}(G)$, and follow the convention of defining $\mathrm{diam}(G) = \infty$, when G is disconnected. In particular, when we say that a graph has finite diameter this implies it is connected.

Let $T = K_1 \log n / A_r = \Theta(n / \log n)$, K_1 is sufficiently large, and $K_1 \ll K$.

Lemma 7. *Let* $u \in S$ *and let* $B = B_{r/2}(u)$. *Then*

$$\mathbf{Pr}\left[\mathrm{diam}(G_n(B)) \geq 2(K_1 + 1) \log n\right] = O(n^{-3})$$

where $G_n(B)$ *is the induced subgraph of* G_n *in* B.

Proof. Given k_0 and N, we consider the following process which generates a sequence of graphs $H_s = (W_s, F_s)$, $s = 1, 2, \ldots, N$. (The meanings of N, k_0 will become apparant soon).

Time Step 1
To initialize the process, we start with H_1 consisting of k_0 isolated vertices $y_1, \ldots, y_{k_0}$.

Time Step $s \geq 1$: We add vertex y_{s+k_0}. We then add $m/4000$ random edges incident with y_{s+k_0} of the form (y_{s+k_0}, w_i) for $i = 1, 2, \ldots, m/4000$. Here each w_i is chosen uniformly from W_s.

The idea is to couple the construction of G_n with the construction of H_N for $N \sim \text{Bi}(n - T, A_r/4)$ and $k_0 = \text{Bi}(T, A_r/4)$ such that **whp** H_N is a subgraph of G_n with vertex set $V_n(B)$. We are going to show that **whp** $\text{diam}(H_N) \leq 2(K_1 + 1)\log n$, and therefore $\text{diam}(G_n(B)) \leq 2(K_1 + 1)\log n$.

To do the coupling we use two counters, t for the steps in G_n and s for the steps in H_N:

- Given G_n, set $s = 0$. Let $W_0 = V_T(B)$. Notice $k_0 = |W_0| \sim \text{Bi}(T, A_r/4)$ and that $k_0 \leq K_1 \log n$ **whp**.
- For every $t > T$.
  - If $x_t \notin B$, do nothing in H_N.
  - If $x_t \in B$, set $s := s + 1$. Set $y_{s+k_0} = x_t$. As we want H_N to be a subgraph of G_n we must choose the neighbors of y_{s+k_0} among the neighbors of G_n. Let A be the set of vertices chosen by x_t in $V_t(B)$. Notice that $|A|$ stochastically dominates $a_t \sim \text{Bi}\left(m, \frac{D_t(B)}{D_t(x_t)}\right)$. If $\frac{D_t(B)}{D_t(x_t)} \geq \frac{1}{100}$, then a_t stochastically dominates $b_t \sim \text{Bi}(m, \frac{1}{100})$ and so **whp** is at least $m/200$. If $\frac{D_t(B)}{D_t(x_t)} < \frac{1}{100}$ we declare failure, but as we see below this is unlikely to happen — see (10). We can assume that $D_t(B) \leq 3mA_r t$ and $k_t = |V_t(B)| \geq A_r t/5$ and so each vertex of B has probability at least $\frac{m}{3mA_r t} \geq \frac{1}{15k_t}$ of being chosen under preferential attachment. Thus, as insightfully observed by Bollobás and Riordan [10] we can legitimately start the addition of x_t in G_t by choosing $\text{Bi}(m, 1/3000)$ random neighbours uniformly in B. Observe that $\text{Bi}(m, 1/3000) \geq m/4000$ **whp**.

Notice that N, the number of times s is increased, is the number of steps for which $x_t \in B$, and so $N \sim \text{Bi}(n - T, A_r/4)$.

Notice also that we have

$$\frac{D_t(B)}{D_t(x_t)} \geq \frac{V_t(B)}{2V_t(B_{2r}(x_t))},$$

and therefore, for $t \geq T$,

$$\mathbf{Pr}\left[\frac{D_t(B)}{D_t(x_t)} \leq \frac{1}{100}\right] \leq \mathbf{Pr}\left[V_t(B) \leq A_r t/6 \text{ or } V_t(B_{2r}(x_t)) \geq 8A_r t\right]$$

$$\leq 2n^{-K_1/8}, \quad (10)$$

where the final inequality follows from Chernoff's bound).

Now we are ready to show that H_N is connected **whp**.

Notice that by Chernoff's bound we get that

$$\mathbf{Pr}\left[\left|k_0 - \frac{K_1}{4}\log n\right| \geq \frac{K_1}{8}\log n\right] \leq 2n^{-K_1/48}$$

and

$$\mathbf{Pr}\left[N \leq \frac{1}{3}(\log n)^2\right] \leq e^{-c(\log n)^2}$$

for some $c > 0$. Therefore, we can assume $\log n \leq k_0 \leq K_1 \log n$ and $N \geq \frac{1}{3}(\log n)^2$.

Let X_s be the number of connected components of H_s. Then

$$X_{s+1} = X_s - Y_s, \qquad X_0 = k_0$$

where Y_s is the number of components (minus one) collapsed into one by y_{s+k_0}. Then

$$\mathbf{Pr}\left[Y_s = 0\right] \leq \sum_{i=1}^{X_s}\left(\frac{c_i}{s+k_0}\right)^{m/4000}$$

where the c_i are the component sizes of H_s. Therefore, if $s < 2K_1\log n$ then, since $m \geq K\log n$, we have

$$\mathbf{Pr}\left[Y_s = 0 \mid X_s \geq 2\right] \leq 2\left(1 - \frac{1}{s+k_0}\right)^{m/4000} \leq e^{-m/(4000(s+k_0))} \leq 1/10.$$

So X_s is stochastically dominated by the random variable $\max(1, k_0 - Z_s)$ where $Z_s \sim \mathrm{Bi}(s, 9/10)$. We get then

$$\mathbf{Pr}\left[X_{2K_1\log n} > 1\right] \leq \mathbf{Pr}\left[Z_{2K_1\log n} < k_0\right] \leq \mathbf{Pr}\left[Z_{2K_1\log n} < K_1\log n\right] \leq n^{-3}.$$

And therefore

$$\mathbf{Pr}\left[H_{2K_1\log n} \text{ is not connected}\right] \leq n^{-3}.$$

Now, to obtain an upper bound on the diameter, we run the process of construction of H_N by rounds. The first round consists of $2K_1\log n$ steps and in each new round we double the size of the graph, i.e. it consists of as many steps as the total number of steps of all the previous rounds. Notice that we have less than $\log n$ rounds in total. Let $\mathcal{A}$ be the event for all $i > 0$ every vertex created in the $(i+1)^{th}$ round is adjacent to a vertex in $H_{2K_1\log n+(i-1)\log n}$, the graph at the end of the ith round.

Conditioning in ρA, every vertex in H_N is at distance at most $\log n$ of $H_{2K_1\log n}$ whose diameter is not greater than $2K_1\log n$. Thus, the diameter of H_N is smaller than $2(K_1 + 1)\log n$.

Now, we have that if v is created in the $(i+1)^{th}$ round,

$$\mathbf{Pr}\left[v \text{ is not adjacent to } H_{2K_1\log n+(i-1)\log n}\right] \leq \left(\frac{1}{2}\right)^m.$$

Therefore
$$\mathbf{Pr}\left[\neg\mathcal{A}\right] \le \left(\frac{1}{2}\right)^m n(\log n) \le \frac{\log n}{n^{K\log 2 - 1}}.$$

$\square$

To finish the proof of connectivity and the diameter, let u, v be two vertices of G_n. Let $C_1, C_2, \ldots, C_M$, $M = O(1/r)$ be a sequence of spherical caps of radius $r/4$ such that u is the center of C_1, v is the center of v and such that the centers of C_i, C_{i+1} are distance $\le r/2$ apart. The intersections of C_i, C_{i+1} have area at least $A_r/40$ and so **whp** each intersection contains a vertex. Using Lemma 7 we deduce that **whp** there is a path from u to v in G_n of size at most $O(\log n/r)$.

References

1. W. Aiello, F. R. K. Chung, and L. Lu, A random graph model for massive graphs, *Proc. of the 32nd Annual ACM Symposium on the Theory of Computing*, (2000) 171–180.
2. W. Aiello, F. R. K. Chung, and L. Lu, Random Evolution in Massive Graphs, *Proc. of IEEE Symposium on Foundations of Computer Science*, (2001) 510–519.
3. R. Albert, A. Barabási, and H. Jeong, Diameter of the world wide web, *Nature* 401 (1999) 103–131.
4. A. Barabasi and R. Albert, Emergence of scaling in random networks, *Science* 286 (1999) 509–512.
5. N. Berger, B. Bollobas, C. Borgs, J. Chayes, and O. Riordan, Degree distribution of the FKP network model, *Proc. of the 30th International Colloquium of Automata, Languages and Programming*, (2003) 725–738.
6. N. Berger, C. Borgs, J. Chayes, R. D'Souza, and R. D. Kleinberg, Competition-induced preferential attachment, preprint.
7. D. Blandford, G. E. Blelloch, and I. Kash, Compact Representations of Separable Graphs, *Proc. of ACM/SIAM Symposium on Discrete Algorithms* (2003) 679–688.
8. B. Bollobás and O. Riordan, Mathematical Results on Scale-free Random Graphs, in *Handbook of Graphs and Networks*, Wiley-VCH, Berlin, 2002.
9. B. Bollobás and O. Riordan, The diameter of a scale-free random graph, *Combinatorica*, 4 (2004) 5–34.
10. B. Bollobás and O. Riordan, Coupling scale free and classical random graphs, preprint.
11. B. Bollobás and O. Riordan and J. Spencer and G. Tusanády, The degree sequence of a scale-free random graph process, *Random Structures and Algorithms* 18 (2001) 279–290.
12. A. Broder, R. Kumar, F. Maghoul, P. Raghavan, S. Rajagopalan, R. Stata, A. Tomkins, and J. Wiener, Graph structure in the web, *Proc. of the 9th Intl. World Wide Web Conference* (2002) 309–320.
13. G. Buckley and D. Osthus, Popularity based random graph models leading to a scale-free degree distribution, *Discrete Mathematics* 282 (2004) 53–68.
14. F.R.K. Chung, L. Lu, and V. Vu, Eigenvalues of random power law graphs, *Annals of Combinatorics* 7 (2003) 21–33.
15. F.R.K. Chung, L. Lu, and V. Vu, The spectra of random graphs with expected degrees, *Proceedings of national Academy of Sciences* 100 (2003) 6313–6318.

16. C. Cooper and A. M. Frieze, A General Model of Undirected Web Graphs, *Random Structures and Algorithms*, 22 (2003) 311–335.
17. E. Drinea, M. Enachescu, and M. Mitzenmacher, *Variations on Random Graph Models for the Web*, Harvard Technical Report TR-06-01 (2001).
18. P. Erdős and A. Rényi, On random graphs I, *Publicationes Mathematicae Debrecen* 6 (1959) 290–297.
19. A. Fabrikant, E. Koutsoupias, and C. H. Papadimitriou, Heuristically Optimized Trade-Offs: A New Paradigm for Power Laws in the Internet, *Proc. of 29th International Colloquium of Automata, Languages and Programming* (2002) .
20. M. Faloutsos, P. Faloutsos, and C. Faloutsos, On Power-law Relationships of the Internet Topology, *ACM SIGCOMM Computer Communication Review* 29 (1999) 251–262.
21. B. Hayes, Graph theory in practice: Part II, *American Scientist* 88 (2000) 104-109.
22. J. M. Kleinberg, R. Kumar, P. Raghavan, S. Rajagopalan, and A. S. Tomkins, The Web as a Graph: Measurements, Models and Methods, *Proc. of the 5th Annual Intl. Conf. on Combinatorics and Computing (COCOON)* (1999).
23. R. Kumar, P. Raghavan, S. Rajagopalan, D. Sivakumar, A. Tomkins, and E. Upfal, Stochastic Models for the Web Graph, *Proc. IEEE Symposium on Foundations of Computer Science* (2000) 57.
24. R. Kumar, P. Raghavan, S. Rajagopalan, D. Sivakumar, A. Tomkins, and E. Upfal, The Web as a Graph, *Proc. 19th ACM SIGACT-SIGMOD-AIGART Symp. Principles of Database Systems (PODS)* (2000) 1–10.
25. R. Kumar, P. Raghavan, S. Rajagopalan, and A. Tomkins, Trawling the Web for emerging cyber-communities, *Computer Networks* 31 (1999) 1481–1493.
26. C. J. H. McDiarmid, Concentration, in *Probabilistic methods in algorithmic discrete mathematics*, (1998) 195-248.
27. M. Mihail and C. H. Papadimitriou, On the Eigenvalue Power Law, *Proc. of the 6th International Workshop on Randomization and Approximation Techniques* (2002) 254–262.
28. M. Mihail, C. H. Papadimitriou, and A. Saberi, On Certain Connectivity Properties of the Internet Topology, *Proc. IEEE Symposium on Foundations of Computer Science* (2003) 28.
29. M. Mitzenmacher, A brief history of generative models for power law and lognormal distributions, preprint.
30. M. D. Penrose, *Random Geometric Graphs*, Oxford University Press (2003).
31. H. A. Simon, On a class of skew distribution functions, *Biometrika* 42 (1955) 425-440.
32. D. J. Watts, *Small Worlds: The Dynamics of Networks between Order and Randomness*, Princeton: Princeton University Press (1999).
33. G. Yule, A mathematical theory of evolution based on the conclusions of Dr. J.C. Willis, *Philosophical Transactions of the Royal Society of London (Series B)* 213 (1925) 21–87.

Traffic-Driven Model of the World Wide Web Graph

Alain Barrat[1], Marc Barthélemy[2], and Alessandro Vespignani[1,3]

[1] Laboratoire de Physique Théorique (UMR du CNRS 8627),
Bâtiment 210, Université de Paris-Sud 91405 Orsay, France
[2] CEA-Centre d'Etudes de Bruyères-le-Châtel, Département de Physique Théorique
et Appliquée BP12, 91680 Bruyères-Le-Châtel, France
[3] School of Informatics, Indiana University, Bloomington, IN 47408, USA

Abstract. We propose a model for the World Wide Web graph that couples the topological growth with the traffic's dynamical evolution. The model is based on a simple traffic-driven dynamics and generates weighted directed graphs exhibiting the statistical properties observed in the Web. In particular, the model yields a non-trivial time evolution of vertices and heavy-tail distributions for the topological and traffic properties. The generated graphs exhibit a complex architecture with a hierarchy of cohesiveness levels similar to those observed in the analysis of real data.

1 Introduction

The World Wide Web (WWW) has evolved into a huge and intricate structure whose understanding represents a major scientific and technological challenge. A fundamental step in this direction is taken with the experimental studies of the WWW graph structure in which vertices and directed edges are identified with web-pages and hyperlinks, respectively. These studies are based on crawlers that explore the WWW connectivity by following the links on each discovered page, thus reconstructing the topological properties of the representative graph. In particular, data gathered in large scale crawls [1, 2, 3, 4, 5] have uncovered the presence of a complex architecture underlying the structure of the WWW graph. A first observation is the *small-world* property [6] which means that the average distance between two vertices (measured by the length of the shortest path) is very small. Another important result is that the WWW exhibits a power-law relationship between the frequency of vertices and their degree, defined as the number of directed edges linking each vertex to its neighbors. This last feature is the signature of a very complex and heterogeneous topology with statistical fluctuations extending over many length scales [1].

These complex topological properties are not exclusive to the WWW and are encountered in a wide range of networked structures belonging to very different domains such as ecology, biology, social and technological systems [7, 8, 9, 10]. The need for general principles explaining the emergence of complex topological

S. Leonardi (Ed.): WAW 2004, LNCS 3243, pp. 56–67, 2004.

features in very diverse systems has led to a wide array of models aimed at capturing various properties of real networks [7, 9, 10], including the WWW. Models do however generally consider only the topological structure and do not take into account the interaction strength –the weight of the link– that characterizes real networks [11, 12, 13, 14, 15, 16]. Interestingly, recent studies of various types of weighted networks [15, 17] have shown additional complex properties such as broad distributions and non-trivial correlations of weights that do not find an explanation just in terms of the underlying topological structure. In the case of the WWW, it has also been recognized that the complexity of the network encompasses not only its topology but also the dynamics of information. Examples of this complexity are navigation patterns, community structures, congestions, and other social phenomena resulting from the users' behavior [18, 19]. In addition, Adamic and Huberman [4] pointed out that the number of users of a web-site is broadly distributed, showing the relevance and heterogeneity of the traffic carried by the WWW.

In this work we propose a simple model for the WWW graph that takes into account the traffic (number of visitors) on the hyper-links and considers the dynamical basic evolution of the system as being driven by the traffic properties of web-pages and hyperlinks. The model also mimics the natural evolution and reinforcements of interactions in the Web by allowing the dynamical evolution of weights during the system growth. The model displays power-law behavior for the different quantities, with non-trivial exponents whose values depend on the model's parameters and which are close to the ones observed empirically. Strikingly, the model recovers a heavy-tailed out-traffic distribution whatever the out-degree distribution. Finally we find non-trivial clustering properties signaling the presence of hierarchy and correlations in the graph architecture, in agreement with what is observed in real data of the WWW.

1.1   Related Works: Existing Models for the Web

It has been realized early that the traditional random graph model, i.e. the Erdös-Renyi paradigm, fails to reproduce the topological features found in the WebGraph such as the broad degree probability distribution, and to provide a model for a dynamical growing network. An important step in the modeling of evolving networks was taken by Barabási et al. [1, 20] who proposed the ingredient of preferential attachment: at each time-step, a new vertex is introduced and connects randomly to already present vertices with a probability proportional to their degree. The combined ingredients of growth and preferential attachment naturally lead to power-law distributed degree. Numerous variations of this model have been formulated [10] to include different features such as re-wiring [21, 22], additional edges, directionality [23, 24], fitness [25] or limited information [26].

A very interesting class of models that considers the main features of the WWW growth has been introduced by Kumar et al. [3] in order to produce a mechanism which does not assume the knowledge of the degree of the existing vertices. Each newly introduced vertex n selects at random an already existing vertex p; for each out-neighbour j of p, n connects to j with a certain probability α; with probability $1 - \alpha$ it connects instead to another randomly chosen node. This model describes the growth process of the WWW as a copy mechanism in which newly arriving web-pages tends to reproduce the hyperlinks of similar web-pages; i.e. the first to which they connect. Interestingly, this model effectively recovers a preferential attachment mechanism without explicitly introducing it.

Other proposals in the WWW modeling include the use of the rank values computed by the PageRank algorithm used in search engines, combined with the preferential attachment ingredient [27], or multilayer models grouping web-pages in different regions [28] in order to obtain bipartite cliques in the network. Finally, recent models include the textual content affinity [29] as the main ingredient of the WWW evolution.

2 Weighted Model of the WWW Graph

2.1 The WWW Graph

The WWW network can be mathematically represented as a directed graph $\mathcal{G} = (V, E)$ where V is the set of nodes which are the web-pages and where E is the set of ordered edges (i, j) which are the *directed* hyperlinks $(i, j = 1, ..., N$ where $N = |V|$ is the size of the network). Each node $i \in V$ has thus an ensemble $\mathcal{V}_{in}(i)$ of pages pointing to i (in-neighbours) and another set $\mathcal{V}_{out}(i)$ of pages directly accessible from i (out-neighbours). The degree $k(i)$ of a node is divided into in-degree $k^{in}(i) = |\mathcal{V}_{in}(i)|$ and out-degree $k^{out}(i) = |\mathcal{V}_{out}(i)|$: $k(i) = k^{in}(i) + k^{out}(i)$. The WWW has also dynamical features in that $\mathcal{G}$ is growing in time, with a continuous creation of new nodes and links. Empirical evidence shows that the distribution of the in-degrees of vertices follows a power-law behavior. Namely, the probability distribution that a node i has in-degree k^{in} behaves as $P(k^{in}) \sim \left(k^{in}\right)^{-\gamma_{in}^k}$, with $\gamma_{in}^k = 2.1 \pm 0.1$ as indicated by the largest data sample [1, 2, 4, 5]. The out-degrees (k^{out}) distribution of web-pages is also broad but with an exponential cut-off, as recent data suggest [2, 5]. While the in-degree represents the sum of all hyper-links coming from the whole WWW and can be in principle as large as the WWW itself, the out-degree is determined by the number of hyper-links present in a single web-page and is thus constrained by obvious physical elements.

2.2 Weights and Strengths

The number of users of any given web-site is also distributed according to a heavy-tail distribution [4]. This fact demonstrates the relevance of considering that every hyper-link has a specific weight that represents the number of users which are navigating on it. The WebGraph $\mathcal{G}(V, E)$ is thus a directed, weighted graph where the directed edges have assigned variables w_{ij} which specify the weight on the edge connecting vertex i to vertex j ($w_{ij} = 0$ if there is no edge pointing from i to j). The standard topological characterization of directed networks is obtained by the analysis of the probability distribution $P(k^{in})$ $[P(k^{out})]$

that a vertex has in-degree k^{in} [out-degree k^{out}]. Similarly, a first characterization of weights is obtained by the distribution $P(w)$ that any given edge has weight w. Along with the degree of a node, a very significative measure of the network properties in terms of the actual weights is obtained by looking at the vertex incoming and outgoing strength defined as [30, 15]

$$s_i^{out} = \sum_{j \in \mathcal{V}_{out}(i)} w_{ij} \; , \; s_i^{in} = \sum_{j \in \mathcal{V}_{in}(i)} w_{ji} \; , \tag{1}$$

and the corresponding distributions $P(s^{in})$ and $P(s^{out})$. The strengths s_i^{in} and s_i^{out} of a node integrate the information about its connectivity and the importance of the weights of its links, and can be considered as the natural generalization of the degree. For the Web the incoming strength represents the actual total traffic arriving at web-page i and is an obvious measure of the popularity and importance of each web-page. The incoming strength obviously increases with the vertex in-degree k_i^{in} and usually displays the power-law behavior $s \sim k^\beta$, with the exponent β depending on the specific network [15].

2.3   The Model

Our goal is to define a model of a growing graph that explicitly takes into account the actual popularity of web-pages as measured by the number of users visiting them. Starting from an initial seed of N_0 pages, a new node (web-page) n is introduced in the system at each time-step and generates m outgoing hyperlinks. In this study, we take m fixed so that the out-degree distribution is a delta function. This choice is motivated by the empirical observation that the distribution of the number of outgoing links is bounded [5] and we have checked that the results do not depend on the precise form of this distribution as long as $P(k^{out}(i) = k)$ decays faster than any power-law as k grows.

The new node n is attached to a node i with probability

$$Prob(n \to i) = \frac{s_i^{in}}{\sum_j s_j^{in}} \tag{2}$$

and the new link $n \to i$ has a weight $w_{ni} \equiv w_0$. This choice relaxes the usual degree preferential attachment and focuses on the popularity—or strength—driven attachment in which new web-pages will connect more likely to web-pages handling larger traffic. This appears to be a plausible mechanism in the WWW and in many other technological networks. For instance, in the Internet new routers connect to other routers with large bandwidth and traffic handling capabilities. In the airport network, new connections (airlines routes) are generally established with airports having a large passenger traffic [15, 31, 32]. The new vertex is assumed to have its own initial incoming strength $s_n^{in} = w_0$ in order to give the vertex an initial non-vanishing probability to be chosen by vertices arriving at later time steps.

The second and determining ingredient of the model consists in considering that a new connection $(n \to i)$ will introduce variations of the traffic across the

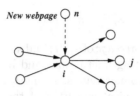

Fig. 1. Illustration of the construction rule. A new web-page n enters the Web and direct a hyper-link to a node i with probability proportional to $s_i^{in}/\sum_j s_j^{in}$. The weight of the new hyper-link is w_0 and the existing traffic on outgoing links of i are modified by a total amount equal to δ_i: $s_i^{out} \rightarrow s_i^{out} + \delta_i$

network. For the sake of simplicity we limit ourselves to the case where the introduction of a new incoming link on node i will trigger only local rearrangements of weights on the existing links $(i \rightarrow j)$ where $j \in \mathcal{V}_{out}(i)$ as

$$w_{ij} \rightarrow w_{ij} + \Delta w_{ij}, \tag{3}$$

where Δw_{ij} is a function of w_{ij} and of the connectivities and strengths of i. In the following we focus on the case where the addition of a new edge with weight w_0 induces a total increase δ_i of the total outgoing traffic and where this perturbation is proportionally distributed among the edges according to their weights [see Fig. 1]

$$\Delta w_{ij} = \delta_i \frac{w_{ij}}{s_i^{out}}. \tag{4}$$

This process reflects the fact that new visitors of a web-page will usually use its hyper-links and thus increase its outgoing traffic. This in turn will increase the popularity of the web-pages pointed by the hyperlinks. In this way the popularity of each page increases not only because of direct link pointing to it but also due to the increased popularity of its in-neighbors. It is possible to consider heterogeneous δ_i distributions depending on the local dynamics and rearrangements specific to each vertex, but for the sake of simplicity we consider the model with $\delta_i = \delta$. We finally note that the quantity w_0 sets the scale of the weights. We can therefore use the rescaled quantities w_{ij}/w_0, s_i/w_0 and δ/w_0, or equivalently set $w_0 = 1$. The model then depends only on the dimensionless parameter δ. The generalization to arbitrary w_0 is simply obtained by replacing δ, w_{ij}, s_i^{out} and s_i^{in} respectively by δ/w_0, w_{ij}/w_0, s_i^{out}/w_0 and s_i^{in}/w_0 in all results.

2.4   Analytical Solution

Starting from an initial seed of N_0 nodes, the network grows with the addition of one node per unit time, until it reaches its final size N. In the model, every new vertex has exactly m outgoing links with the same weight $w_0 = 1$. During the growth process this symmetry is conserved and at all times we have $s_i^{out} = m w_{ij}$. Indeed, each new incoming link generates a traffic reinforcement $\Delta w_{ij} = \delta/m$, so that $w_{ij} = w_0 + k_i^{in}\delta/m$ is independent from j and

$$s_i^{out} = m + \delta k_i^{in} . \tag{5}$$

The time evolution of the *average* of $s_i^{in}(t)$ and $k_i^{in}(t)$ of the i-th vertex at time t can be obtained by neglecting fluctuations and by relying on the continuous approximation that treats connectivities, strengths, and time t as continuous variables [7, 9, 10]. The dynamical evolution of the in-strength of a node i is given by the evolution equation

$$\frac{ds_i^{in}}{dt} = m\frac{s_i^{in}}{\sum_l s_l^{in}} + \sum_{j \in V_{in}(i)} m\frac{s_j^{in}}{\sum_l s_l^{in}}\delta\frac{1}{m} , \tag{6}$$

with initial condition $s_i^{in}(t = i) = 1$. This equation states that the incoming strength of a vertex i can only increase if a new hyper-link connects directly to i (first term) or to a neighbor vertex $j \in V_{in}(i)$, thus inducing a reinforcement δ/m on the existing in-link (second term). Both terms are weighted by the probability that the new vertex establishes a hyperlink with the corresponding existing vertex. Analogously, we can write the evolution equation for the in-degree k_i^{in} that evolves only if the new link connects directly to i:

$$\frac{dk_i^{in}}{dt} = m\frac{s_i^{in}}{\sum_l s_l^{in}} . \tag{7}$$

Finally, the out-degree is constant ($k_i^{out} = m$) by construction.

The above equations can be written more explicitly by noting that the addition of each new vertex and its m out-links, increase the total in-strength of the graph by the constant quantities $1 + m + m\delta$ yielding at large times $\sum_{l=1}^{t} s_l^{in} = m\left(1 + \frac{1}{m} + \delta\right)t$. By inserting this relation in the evolution equations (6) and (7) we obtain

$$\frac{ds_i^{in}}{dt} = \frac{1}{\delta + 1 + \frac{1}{m}}\left(\frac{s_i^{in}}{t} + \frac{\delta}{mt}\sum_{j \in V_{in}(i)} s_j^{in}\right) \quad\text{and}\quad \frac{dk_i^{in}}{dt} = \frac{1}{\delta + 1 + \frac{1}{m}}\frac{s_i^{in}}{t} . \tag{8}$$

These equations cannot be explicitly solved because of the term $\sum_{j \in V_{in}(i)} s_j^{in}$ which introduces a coupling of the in-strength of different vertices. The structure of the equations and previous studies of similar undirected models [31, 32] suggest to consider the *Ansatz* $s_i^{in} = Ak_i^{in}$ in order to obtain an explicit solution. Using (5), and $w_{ji} = s_j^{out}/m$, we can write

$$s_i^{in} = \sum_{j \in V_{in}(i)} w_{ji} = k_i^{in} + \sum_{j \in V_{in}(i)} \frac{\delta}{m}k_j^{in}, \tag{9}$$

and the Ansatz $s_i^{in} = Ak_i^{in}$ yields

$$\sum_{j \in V_{in}(i)} s_j^{in} = \frac{m}{\delta}(A - 1)s_i^{in} . \tag{10}$$

This allows to have a closed equation for s_i^{in} whose solution is

$$s_i^{in}(t) = \left(\frac{t}{i}\right)^{\theta} , \quad \text{with} \quad \theta = \frac{A}{\delta + 1 + 1/m} \tag{11}$$

and $k_i^{in}(t) = s_i^{in}(t)/A$, satisfying the proposed Ansatz. The fact that vertices are added at a constant rate implies that the probability distribution of s_i^{in} is given by [10, 31, 32]

$$P(s^{in}, t) = \frac{1}{t + N_0} \int_0^t \delta(s^{in} - s_i^{in}(t))di, \tag{12}$$

where $\delta(x)$ is the Dirac delta function. By solving the above integral and considering the infinite size limit $t \to \infty$ we obtain

$$P(s^{in}(i) = s) \sim s^{-\gamma_{in}^s} , \quad \text{with} \quad \gamma_{in}^s = 1 + \frac{1}{\theta} \tag{13}$$

The quantities s_i^{in}, k_i^{in} and s_i^{out} are thus here proportional, so that their probability distributions are given by power-laws with the same exponent $\gamma_{in}^s = \gamma_{out}^s = \gamma_{in}^k$. The explicit value of the exponents depends on θ which itself is a function of the proportionality constant A. In order to find an explicit value of A we use the approximation that on average the total in-weight will be proportional to the number of in-links times the average weight in the graph $< w >= \frac{1}{tm} \sum_l s_l^{out} = (\delta + 1)$. At this level of approximation, the exponent θ varies between $m/(m+1)$ and 1 and the power-law exponent thus varies between 2 ($\delta \to \infty$) and $2 + 1/m$ ($\delta = 0$). This result points out that the model predicts an exponent $\gamma_{in}^k \simeq 2$ for reasonable values of the out-degree, in agreement with the empirical findings.

3   Numerical Simulations

Along with the previous analytical discussion we have performed numerical simulations of the presented graph model in order to investigate its topological properties with a direct statistical analysis.

3.1   Degree and Strength Distributions

As a first test of the analytical framework we confirm numerically that s^{in}, k^{in}, s^{out} are indeed proportional and grow as power-laws of time during the construction of the network [see Fig. 2]. The measure of the proportionality factor A between s^{in} and k^{in} allows to compute the exponents θ and γ, which are satisfactorily close to the observed results and to the theoretical predictions obtained with the approximation $A \approx < w >$. Fig. 2 shows the probability distributions of k^{in}, w, s^{in}, and s^{out} for $\delta = 0.5$. All these quantities are broadly distributed according to power-laws with the same exponent. It is also important to stress that the out-traffic is broadly distributed even if the out-degree is not.

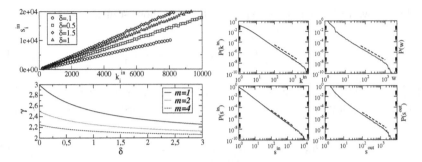

Fig. 2. Top left: illustration of the proportionality between s^{in} and k^{in} for various values of δ. Bottom left: theoretical approximate estimate of the exponent $\gamma_{in}^s = \gamma_{out}^s = \gamma_{in}^k$ vs. δ for various values of m. Right: Probability distributions of k^{in}, w, s^{in}, s^{out} for $\delta = 0.5$, $m = 2$ and $N = 10^5$. The dashed lines correspond to a power law with exponent $\gamma = 2.17$ obtained by measuring first the slope A of s^{in} vs. k^{in} and then using (11) and (13) to compute γ

3.2 Clustering and Hierarchies

Along with the vertices hierarchy imposed by the strength distributions the WWW displays also a non-trivial architecture which reflects the existence of well defined groups or communities and of other administrative and social factors. In order to uncover these structures a first characterization can be done at the level of the undirected graph representation. In this graph, the degree of a node is the sum of its in- and out-degree ($k_i = k_i^{in} + k_i^{out}$) and the total strength is the sum of its in- and out-strength ($s_i = s_i^{in} + s_i^{out}$). A very useful quantity is then the clustering coefficient c_i that measures the local group cohesiveness and is defined for any vertex i as the fraction of connected neighbors couples of i [6]. The average clustering coefficient $C = N^{-1} \sum_i c_i$ thus expresses the statistical level of cohesiveness by measuring the average density of interconnected vertex triplets in the network. Further information can be gathered by inspecting the average clustering coefficient $C(k)$ restricted to vertices with degree k [33, 35]

$$C(k) = \frac{1}{N_k} \sum_{i/k_i=k} c_i \,, \tag{14}$$

where N_k is the number of vertices with degree k. In real WWW data, it has been observed that the k spectrum of the clustering coefficient has a highly non-trivial behavior with a power-law decay as a function of k, signaling a hierarchy in which low degree vertices belong generally to well interconnected communities (high clustering coefficient) while hubs connect many vertices that are not directly connected (small clustering coefficient) [34, 35].

We show in Fig. 3 how the clustering coefficient $C(k)$ of the model we propose increases with δ. We obtain for $C(k)$ a decreasing function of the degree k, in agreement with real data observation. In addition, the range of variations spans

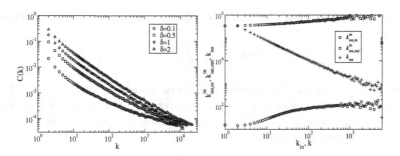

Fig. 3. Left: Clustering coefficient $C(k)$, for various values of the parameter δ. Here $m = 2$ and $N = 10^5$. The clustering increases with δ. Right: Correlations between degrees of neighbouring vertices as measured by $k_{nn}(k)$ (crosses), $k_{nn,in}^{in}(k^{in})$ (circles) and $k_{nn,out}^{in}(k^{in})$ (squares); $m = 2$, $\delta = 0.5$ and $N = 10^5$

several orders of magnitude indicating a continuum hierarchy of cohesiveness levels as in the analysis of Ref. [35].

Interestingly, the clustering spectrum can be qualitatively understood by considering the dynamical process leading to the formation of the network. Indeed, vertices with large connectivities and strengths are the ones that entered the system at the early times, as shown by (11). This process naturally builds up a hierarchy among the nodes, the "older" vertices having larger connectivities. Newly arriving vertices attach to pre-existing vertices with large strength which on their turn are reinforced by the rearrangement of the weights, as well as their neighbours. The increase of C with δ is directly related to this mechanism: each time the extremity i of an edge is chosen by preferential attachment, i and its m out-neighbours are reinforced, thus increasing the probability that, at a following step, a new node connects to both i *and* one of its m neighbours, forming a trangle. Triangles will therefore typically be made of two "old" nodes and a "young" one. This explains why $C(k)$ is large for smaller degree and also why $C(k)$ increases faster for smaller k when δ increases. In contrast, increasing δ does not affect much the "older" nodes which implies that for large degrees, the clustering coefficient is not significantly affected by variation of δ. These properties can be simply reconciled with a real dynamical feature of the web. Newly arriving web-pages will likely point to well known pages that, on their turn, are mutually pointing each other with high probability, thus generating a large clustering coefficient. On the contrary, well known and long established web-pages are pointed by many less popular and new web-pages with no hyper-links among them. This finally results in a small clustering coefficient for well known high degree pages. The structure of the clustering coefficient is therefore the mathematical expression of the structure imposed to the web by its commu-nity structure that generally forms cliques of "fans" web-pages pointing to sets of interconnected "authority" web-pages.

Another important source of information about the network structural orga-nization lies in the correlations of the connectivities of neighboring vertices [36].

Correlations can be probed by inspecting the average degree of nearest neighbor of a vertex i

$$k_{nn,i} = \frac{1}{k_i} \sum_{j \in \mathcal{V}(i)} k_j , \qquad (15)$$

where the sum runs on the nearest neighbors vertices of each vertex i. From this quantity a convenient measure to investigate the behavior of the degree correlation function is obtained by the average degree of the nearest neighbors, $k_{nn}(k)$, for vertices of degree k [33]

$$k_{nn}(k) = \frac{1}{N_k} \sum_{i/k_i=k} k_{nn,i}. \qquad (16)$$

This last quantity is related to the correlations between the degree of connected vertices since on the average it can be expressed as $k_{nn}(k) = \sum_{k'} k' P(k'|k)$, where $P(k'|k)$ is the conditional probability that, given a vertex with degree k, it is connected to a vertex with degree k'. If degrees of neighboring vertices are uncorrelated, $P(k'|k)$ is only a function of k', i.e. each link points to a vertex of a given degree with the same probability independently on the degree of the emanating vertex, and thus $k_{nn}(k)$ is a constant. When correlations are present, two main classes of possible correlations have been identified: *Assortative* behavior if $k_{nn}(k)$ increases with k, which indicates that large degree vertices are preferentially connected with other large degree vertices, and *disassortative* if $k_{nn}(k)$ decreases with k [37].

In the case of the WWW, however, the study of additional correlation function is naturally introduced by the directed nature of the graph. We focus on the most significant, the in-degree of vertices that in our model is a first measure of their popularity. As for the undirected correlation, we can study the average in-degree of in-neighbours :

$$k_{nn,in}^{in}(i) = \frac{1}{k^{in}(i)} \sum_{j \in \mathcal{V}_{in}(i)} k^{in}(j) . \qquad (17)$$

This quantity measures the average in-degree of the in-neighbours of i, i.e. if the pages pointing to a given page i are popular on their turn. Moreover, relevant information comes also from

$$k_{nn,out}^{in}(i) = \frac{1}{k^{out}(i)} \sum_{j \in \mathcal{V}_{out}(i)} k^{in}(j) , \qquad (18)$$

which measures the average in-degree of the out-neighbours of i, i.e. the popularity of the pages to which page i is pointing. Finally, in both cases it is possible to look at the average of this quantity for group of vertices with in-degree k_i^{in} in order to study the eventual assortative or disassortative behavior.

In Fig. 3 we report the spectrum of $k_{nn}(k)$, $k_{nn,in}^{in}(k^{in})$ and $k_{nn,out}^{in}(k^{in})$ in graphs generated with the present weighted model. The undirected correlations display a strong disassortative behaviour with k_{nn} decreasing as a power-law.

This is a common feature of most technological networks which present a hierarchical structure in which small vertices connect to hubs. The model defined here exhibits spontaneously the hierarchical construction that is observed in real technological networks and the WWW. In contrast, both $k_{nn,in}^{in}(k^{in})$ and $k_{nn,out}^{in}(k^{in})$ show a rather flat behavior signaling an absence of strong correlations. This indicates a lack of correlations in the popularity, as measured by the in-degree. The absence of correlations in the behaviour of $k_{nn,out}^{in}(k^{in})$ is a realistic feature since in the real WWW, vertices tend to point to popular vertices independently of their in-degree. We also note that $k_{nn,out}^{in}(k^{in}) \gg k_{nn,in}^{in}(k^{in})$, a signature of the fact that the average in-degree of pointed vertices is much higher than the average in-degree of pointing vertices. This result also is a reasonable feature of the real WWW since the average popularity of webpages to which any vertex is pointing is on average larger than the popularity of pointing webpages that include also the non-popular ones.

Finally, we would like to stress that in our model the degree correlations are to a certain extent a measure of popularity correlations and more refined measurements will be provided by the correlations among the actual popularity as measured by the in-strength of vertices. We defer the detailed analysis of these properties to a future publication, but at this stage, it is clear that an empirical analysis of the hyperlinks traffic is strongly needed in order to discuss in detail the WWW architecture.

4 Conclusion

We have presented a model for the WWW that considers the interplay between the topology and the traffic dynamical evolution when new web-pages and hyperlinks are created. This simple mechanism produces a non trivial complex and scale-free behavior depending on the physical parameter δ that controls the local microscopic dynamics. We believe that the present model might provide a general starting point for the realistic modeling of the Web by taking into account the coupling of its two main complex features, its topology and its traffic.

Acknowledgments. A.B and A. V. are partially funded by the European Commission - Fet Open project COSIN IST-2001-33555 and contract 001907 (DELIS).

References

1. A.-L. Barabasi and R. Albert, Science **286**, 509 (1999)
2. A. Broder, R. Kumar, F. Maghoul, P. Raghavan, S. Rajagopalan, S. Stata, A. Tomkins and J. Wiener, *Graph structure in the Web*, in *Proceedings of the 9th WWW conference* (2000).
3. R. Kumar, et al, *Stochastic models for the Web graph*, in 'Proceedings of the 41th IEEE Symposium on Foundations of Computer Science (FOCS)', pp. 57-65 (2000).

4. L.A. Adamic and B.A. Huberman, in 'Communications of the ACM', **44**, ACM Press New York, NY, USA, pp. 55-60 (2001).
5. L. Laura, S. Leonardi, S. Millozzi, U. Meyer, and J.F. Sibeyn, 'Algorithms and experiments for the Webgraph'. European Symposium on Algorithms (2003).
6. Watts D. J. and Strogatz S. H., Nature **393**, 440 (1998).
7. R. Albert and A.-L. Barabási, Rev. Mod. Phys. **74**, 47 (2000).
8. L.A.N. Amaral, A. Scala, M. Barthélemy, and H.E. Stanley, Proc. Natl. Acad. Sci. *USA* **97**, 11149 (2000).
9. S. N. Dorogovtsev and J. F. F. Mendes, *Evolution of networks: From biological nets to the Internet and WWW* (Oxford University Press, Oxford, 2003).
10. R. Pastor-Satorras and A. Vespignani, *Evolution and structure of the Internet: A statistical physics approach* (Cambridge University Press, Cambridge, 2004).
11. M. Granovetter, American Journal of Sociology, **78** (6) 1360-1380 (1973).
12. M. E. J. Newman, Phys. Rev. E **65**, 016131 (2001); M. E. J. Newman, Phys. Rev. E **65**, 016132 (2001).
13. A.-L. Barabasi, H. Jeong, R. Ravasz, Z. Neda, T. Vicsek, and A. Schubert, Physica A **311**, 590 (2002).
14. R. Guimera, S. Mossa, A. Turtschi, and L.A.N. Amaral, submitted (2003).
15. A. Barrat, M. Barthélemy, R. Pastor-Satorras, and A. Vespignani, Proc. Natl. Acad. Sci. (USA), **101**, 3747 (2004).
16. C. Quince, P.G. Higgs, and A.J. McKane, arXiv:q-bio.PE/0402014 (2004).
17. D. Garlaschelli et al, submitted (2003). Cond-mat-0310503.
18. B.A. Huberman and R. Lukose, Science **277**, 535 (1997).
19. B.A. Huberman, P. Pirolli, J. Pitkow, and R. Lukose, Science **280**, 95 (1998).
20. A.-L. Barabasi, R. Albert, and H. Jeong, Physica A **281**, 69 (2000)
21. B. Tadic, Physica A **293**, 273 (2001).
22. P.L. Krapivsky, G. Rodgers, and S. Redner, Phys. Rev. Lett. **86**, 5401 (2001).
23. S.N. Dorogovtsev and J.F.F. Mendes, Europhys. Lett. **52**, 33 (2000).
24. C. Cooper and A. Frieze, *A General Model of Undirected Web Graphs*, Proceedings of the 9th Annual European Symposium on Algorithms, Lecture Notes in Computer Science 2161, p. 500-511 (2001).
25. G. Bianconi and A.-L. Barabasi, Europhys. Lett. **54**, 436 (2001).
26. S. Mossa, M. Barthélemy, H.E. Stanley, and L.A.N. Amaral, Phys. Rev. Lett. **88**, 138701 (2002).
27. G. Pandurangan, P. Raghavan and E. Upfal, *Using PageRank to Characterize Web Structure*, In 'Proceedings of the 8th Annual International Computing and Combinatorics Conference (COCOON)', LNCS 2387, p. 330-339 (2002).
28. L. Laura, S. Leonardi, G. Caldarelli and P. De Los Rios, *A multi-layer model for the webgraph*, in 'On-line proceedings of the 2nd international workshop on web dynamics' (2002).
29. F. Menczer, Proc. Natl. Acad. Sci. *USA* **99**, 14014 (2002).
30. S.H. Yook, H. Jeong, A.-L. Barabasi, and Y. Tu, Phys. Rev. Lett. **86**, 5835 (2001).
31. A. Barrat, M. Barthélemy, and A. Vespignani, Phys. Rev. Lett. **92**, 228701 (2004).
32. A. Barrat, M. Barthélemy, and A. Vespignani, cond-mat/0406238.
33. A. Vázquez, R. Pastor-Satorras and A. Vespignani, *Phys. Rev. E* **65**, 066130 (2002).
34. J.-P. Eckmann and E. Moses, Proc. Natl. Acad. Sci. (USA), **99**, 5825 (2002).
35. Ravasz, E. & Barabási, A.-L. *Phys. Rev. E* **67**, 026112 (2003).
36. R. Pastor-Satorras, A. Vázquez and A. Vespignani, *Phys. Rev. Lett.* **87**, 258701 (2001).
37. M. E. J. Newman, *Phys. Rev. Lett.* **89**, 208701 (2002).

On Reshaping of Clustering Coefficients in Degree-Based Topology Generators

Xiafeng Li, Derek Leonard, and Dmitri Loguinov

Texas A&M University, College Station, TX 77843, USA
{xiafeng, dleonard, dmitri}@cs.tamu.edu

Abstract. Recent work has shown that the Internet exhibits a power-law node degree distribution and *high* clustering. Considering that many existing degree-based Internet topology generators do not achieve this level of clustering, we propose a randomized algorithm that increases the clustering coefficients of graphs produced by these generators. Simulation results confirm that our algorithm makes the graphs produced by existing generators match clustering properties of the Internet topology.

1 Introduction

Many studies [5], [7], [10] examine the properties of the Internet connectivity graph and attempt to understand its evolution. Results in [10], [19] demonstrate that both AS-level and router-level Internet topology exhibits three important characteristics: *power-law degree distribution, small diameter*, and *high clustering*. To satisfy these properties, many degree-based generators have been proposed to model the Internet graph [5], [7], [10], [21], [33]. Although most of them can achieve the necessary degree distributions and diameter, they are often not successful in producing high levels of clustering. As shown in [10], the graphs produced by the existing Internet generators always exhibit much lower clustering than the Internet graph. In order to make these graphs better match the Internet, we propose a randomized algorithm that increases the clustering of synthetic graphs while preserving their power-law degree distributions and small diameters.

Considering that the algorithm spends most of its time on computing clustering coefficients during each iteration, we use sampling theory and let the method utilize approximate values instead of computing them exactly, which reduces the time complexity of the algorithm from $\Theta(mn)$ to $\Theta(m)$, where m is the number of edges and n is the number of nodes. This paper makes two contributions. First, it proposes a new and practical method to estimate the clustering of massive graphs by using sampling theory. Second, it offers a solution to the low-clustering problem of existing Internet topology generators.

The remainder of the paper is organized as follows. First, we review the important properties of the Internet topology and classify existing generators in section 2. Then, we abstract the low clustering problem in these generators and propose our algorithmic solution in section 3. Following that, the analysis of the

S. Leonardi (Ed.): WAW 2004, LNCS 3243, pp. 68–79, 2004.

algorithm and the corresponding simulation results are given in section 4. We conclude the paper in section 5.

2 Background

In this section, we first review properties of the Internet topology and then discuss existing degree-based Internet topology generators.

2.1 Internet Properties

It is important to study the properties of the Internet topology, because they not only affect the deployment of Internet services [18], [20], but also impact the performance of existing Internet protocols [22], [23], [30]. Past work has showed that the Internet AS-level graph exhibits the following three properties.

The first property of the Internet is related to its degree distribution. In 1999, Faloutsos et al. [19] observed that the CDF of node degree X (both AS-level and router-level) follows a power-law (Pareto) distribution:

$$P(X \leq d) = 1 - cd^{-\alpha} \ , \tag{1}$$

where $\alpha \approx 1.2$ is the shape parameter and c is the scale parameter.

The second property of the Internet topology is its high clustering. In 2002, Bu et al. [10] pointed out that the Internet is highly clustered. The paper showed that the *clustering coefficient* [32], which is a metric of the probability that two neighbors share a common third neighbor, of the Internet topology is much higher than that of random graphs.

The third property is that the Internet graph has a small diameter. Bu et al. [10] showed that the *average shortest path length* between each pair of nodes at the AS-level is very small and is close to that of random graphs.

To reproduce these properties, many generators have been proposed to model the Internet [2], [3], [4], [9], [10], [33]. As most of them seek to satisfy the first property (the power-law degree distribution), they are also called *degree-based* generators. Next, we review several widely-used generators and classify them into two types.

2.2 Existing Internet Topology Generators

Despite the various implementations, existing degree-based Internet generators can be categorized into two collections: *evolving* and *non-evolving* generators.

Non-evolving methods do not model the evolution of the Internet and produce graphs with a given fixed number of nodes n. GED [14], [26] and PLRG [5] are classical examples belonging to this collection. In GED (Given Expected Degree) [14], [26], [27], [28], n pre-assigned weights $(w_1, \ldots, w_n)$ are drawn from a Pareto distribution. Edge (i, j) exists with probability p_{ij}:

$$p_{ij} = \frac{w_i w_j}{\sum_{k=1}^{n} w_k} \ . \tag{2}$$

To make p_{ij} less than or equal to 1, the pre-assigned degree sequence is assumed to satisfy the following condition [14], [26]:

$$w_{max}^2 \le \sum_{k=1}^{n} w_k \ , \tag{3}$$

where $w_{max} = \max_{i=1}^{n}\{w_i\}$.

To relax this assumption, the PLRG model [5] is proposed. In PLRG, a power-law weight sequence $\{w_i\}$ is first pre-assigned to n nodes. After that, w_i virtual copies of node i are produced and are randomly selected to form links with equal probability. Because of their simplicity, PLRG and GED are good theoretical models for complex networks and many analytical results [5], [14], [15], [26] are based on them. However, they exhibit much lower clustering than the real Internet.

Unlike non-evolving generators, evolving methods focus on modelling the evolution of the Internet topology. In 1999, Barabasi *et al.* [7] proposed the *BA* model, in which the graph evolves by adding new nodes that attach to existing nodes with a so-called *linear preferential* probability $\prod(d_i)$:

$$\prod(d_i) = \frac{d_i(t)}{\sum_j d_j(t)} \ , \tag{4}$$

where d_i is the degree of node i at time t. As shown in [7], the BA model is scale-free and generates graphs with a power-law degree distribution. However, the shape parameter of the power-law function does not match that of the Internet. Moreover, the clustering of a BA-generated graph is much smaller than that of the Internet, which limits the model's usefulness in simulating the Internet.

In order to make the BA model match the Internet more accurately, many BA-like models [1], [6], [10], [31], [33] have been proposed. One of the most successful methods is GLP, which extends the BA model by using a Generalized Linear Preference probability:

$$\prod(d_i) = \frac{d_i(t) - \beta}{\sum_j(d_j(t) - \beta)} \ , \tag{5}$$

where $\beta \in (-\infty, 1]$ is a tunable parameter. Simulation results in [10] show that GLP not only improves the power-law distribution of BA, but also has clustering as high as 0.35. However, compared with the value of 0.45 in the Internet, GLP also needs to increase its clustering [10].

3 Clustering Problem and Algorithmic Solution

Recall that the Internet structure exhibits high clustering, but most existing Internet generators fail to imitate this property. To confirm the clustering inconsistency between the Internet and its generators, we next compare the clustering evolution of the Internet topology with that of its generators.

3.1   Clustering Evolution

To obtain the clustering evolution of the Internet graph, we examined AS (autonomous system) BGP logs from the National Laboratory for Applied Network Research (NLANR) [29], which has been recording a snapshot of the Internet topology every day since 1997. Obvious errors (i.e., duplicate links between some nodes and disconnected components) in a few graphs indicate that part of the data is unreliable and must be removed before the analysis. The assumption on which we filter the data is that the Internet topology must be a connected, simple graph and the number of ASes should increase over time. Based on this assumption, we removed self-loops, merged duplicate links, and discarded graphs that were disconnected or had much fewer ASes than in previously recorded graphs. After preprocessing the data in this manner, we obtained 253 snapshots of the Internet that correspond to the AS-level Internet topology from November 1997 to January 2000.

The clustering coefficients of the obtained graphs are plotted in Figure 1 (left), where γ increases from 0.34 in 1997 to 0.42 in 2000 at a slow, but steady rate. At the same time, the number of nodes in the system also grows as shown on the right side of the figure.

To compare the clustering evolution of the Internet with that of its generators, we simulate the clustering evolutions of GED, PLRG, BA and GLP in

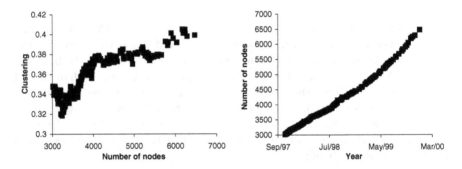

Fig. 1. Evolution of clustering (left) and the number of nodes (right) in the Internet

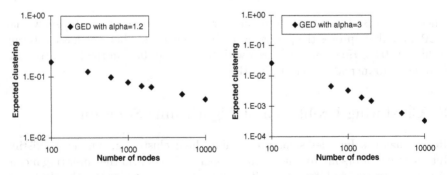

Fig. 2. Clustering evolution of GED with $\alpha = 1.2$ (left) and $\alpha = 3$ (right)

Figures 2, 3, 4 and 5. In the simulations, the number of nodes n increases while other parameters in these models are fixed. Comparing Figure 1(left) with Figures 2, 3, 4 and 5, we conclude that the graphs produced by those generators should be altered to exhibit clustering as high as that of the Internet. Consid-

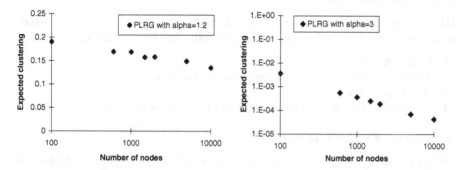

Fig. 3. Clustering evolution of PLRG with $\alpha = 1.2$ (left) and $\alpha = 3$ (right)

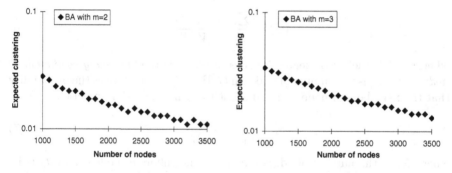

Fig. 4. Clustering evolution of BA with $m = 2$ (left) and $m = 3$ (right)

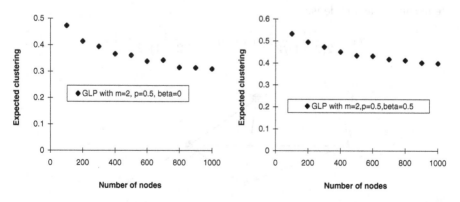

Fig. 5. Clustering evolution of GLP with $m = 2, p = 0.5, \beta = 0$ (left) and $m = 2, p = 0.5, \beta = 0.5$ (right)

ering that the power-law degree sequence and low diameter are also necessary properties in these graphs, we need to keep these properties unchanged while increasing clustering of the graph. The details of the problem can be described as follows.

3.2 Clustering Problem

Given a *connected* graph G and a target clustering value γ_T, rewire G's edges and produce a new graph G' satisfying the following four conditions:

1. G' is connected.
2. The degree sequence in G is the same as that in G'.
3. G' has low diameter.
4. Clustering of G' is larger than or equal to γ_T, which means $\gamma(G') \geq \gamma_T$.

3.3 Clustering and Triangles

To better understand this problem, we next review the concept of clustering and reveal how it relates to triangles.

The *clustering coefficient of a graph* $G(V, E)$, denoted by $\gamma(G)$, is the average clustering coefficient of each node with degree larger than 1:

$$\gamma(G) = \frac{\sum_{v \in V - V^{(1)}} \gamma_v}{|V| - |V^{(1)}|}, \tag{6}$$

where $V^{(1)}$ is the set of degree-1 nodes in G, γ_v is the *clustering coefficient of node* v and $|V|$ is the number of nodes in G. Here, γ_v characterizes the probability that the neighbors of node v are adjacent to each other. More precisely,

$$\gamma_v = \frac{N_v}{d_v(d_v - 1)/2}, \tag{7}$$

where N_v is the number of edges among the neighbors of node v and d_v is its degree. An example of computing clustering is shown in Figure 6, where graph G contains five nodes A, B, C, D, and E. According to the definition, the clustering coefficient of each node is:

$$\gamma_A = \frac{1}{2(2-1)/2} = 1, \quad \gamma_B = \frac{1}{2(2-1)/2} = 1,$$

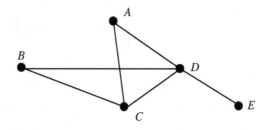

Fig. 6. Clustering in graph G

$$\gamma_C = \frac{2}{3(3-1)/2} = \frac{2}{3}, \quad \gamma_D = \frac{2}{4(4-1)/2} = \frac{1}{3}.$$

And the clustering coefficient of graph G is

$$\gamma(G) = \frac{\gamma_A + \gamma_B + \gamma_C + \gamma_D}{4} = \frac{3}{4}. \tag{8}$$

Note that N_v in (7) is essentially the number of triangles containing node v. For example, in Figure 6, $N_A = 1$ because it is in one triangle (DAC); $N_C = 2$ because it is contained by two triangles (DCB) and (DCA). Therefore, for any node with fixed degree d, the more triangles it includes, the higher clustering it has. Also note that the clustering of a graph is the average of each node's clustering. Intuitively, increasing the number of triangles is a promising way to increase the clustering of a graph.

3.4   Our Algorithm

The key idea of our algorithm is to increase the number of triangles for each node. According to condition 2 in the clustering problem, the degree of each node v should not be changed, which indicates that increasing N_v in (7) will increase the clustering of node v. Therefore, rewiring the links in G to produce more triangles for each node will increase the clustering of the whole graph.

To better describe our algorithm, we first give the definition of *unsatisfied* and *satisfied* nodes as follows.

Definition 1. *A node $v' \in G'$ is unsatisfied if $d_v > d'_v$, $v \in G$. Otherwise, v' is satisfied.*

For example, in Figure 6, if we remove edge (A, C) from the graph, nodes A and C are unsatisfied because their degree decreases. This simple definition facilitates the explanation of our algorithm, which can be separated into four steps. The first step finds all triangles in G and *marks* the corresponding links in these triangles. Then, it randomly picks a node w and searches for k-length $(k \geq 4)$ loops starting from node w. At each time when such a loop is found, our algorithm randomly breaks an *unmarked* link (u, v) from that loop and marks nodes u, v *unsatisfied*. In the third step, the algorithm adds links between any pair of *unsatisfied* nodes so that at least one new triangle is generated. This step is repeated until the clustering of current graph is larger than γ_T or there are no *unsatisfied* nodes remaining. Finally, if the current clustering $\gamma_c(G)$ is larger than γ_T, the algorithm randomly adds links between *unsatisfied* nodes and outputs G'. Otherwise, the method loops back to step two.

In step four, the time complexity of computing current clustering $\gamma_c(G)$ is $\Theta(nm)$, while step 1 to step 3 will only cost $\Theta(m)$. Obviously, reducing the time complexity of computing $\gamma_c(G)$ will improve the performance of our algorithm. Therefore, in step four we randomly sample s nodes and approximate the clustering of the whole graph by the average clustering of the sampled nodes. By applying this randomized sampling technique, the time complexity of step four

Input: a connected, power-law graph G and target clustering γ_T.
Output: a connected, power-law graph G', such that $\gamma(G') \geq \gamma_T$.

Copy graph G to graph G'.
Use *BFS* to find all triangles in G' and mark all corresponding edges in the triangles.
Randomly sample s nodes and compute $\gamma_s(G')$, which is the average clustering coefficient of the s nodes.
While $\gamma_s(G') < \gamma_T$ *do*
    Randomly pick a node w in G'.
    Start from w and apply *BFS* to find all k-cycles ($k \geq 4$) in the graph.
    If there are no such cycles, output *Fail.*
    Else For each k-cycle l, randomly break its unmarked edge (u, v).
    While there exist at least two unsatisfied nodes *do*
        If there exist unsatisfied nodes s and t such that edge $(s, t) \notin G'$
        AND connecting s and t creates at least one triangle *do*
            Connect s and t
        Else if there exist unsatisfied nodes u, v, and w such that there are no edges
among them and $d_u - d'_u \geq 2$, $d_v - d'_v \geq 2$, and $d_w - d'_w \geq 2$;
        AND connecting nodes u, v and w creates one new triangle
            connects links $(u, v), (u, w)$ and (v, w).
        Else
            break the while loop;
        Endif
    EndWhile;
    Randomly sample s nodes and compute $\gamma_s(G')$, the average clustering of the s nodes.
EndWhile;
Randomly connect unsatisfied nodes and output G'.

Fig. 7. Algorithm to increase clustering coefficients of random graphs

is reduced to $\Theta(sm) = \Theta(m)$. A detailed description of the algorithm is shown in Figure 7.

4   Analysis of the Algorithm

There are two issues that must be explored in order to show that our algorithm is effective. We first study through simulations the effect of running our algorithm on random graphs created by several of the methods mentioned before. Since an approximation of clustering in the graph is used in the algorithm, we then analytically determine how accurate these approximations can become.

To show that our algorithm indeed increases the clustering in a wide range of random graphs, we ran sample graphs created by BA, GED, and PLRG through the algorithm. The results are displayed in Fig. 8 and Fig. 9. In each case the graph contains 1000 nodes. For the graph generated by BA, $m = 2$. In the case of both PLRG and GED, $\alpha = 1.5$. Note that duplicate links and self-loops were removed from these graphs before we ran the algorithm. As shown in the

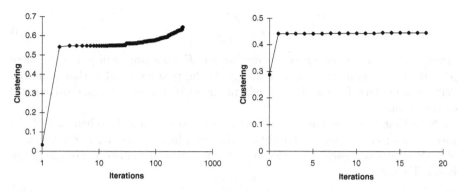

Fig. 8. Increase in clustering for BA graph of 1000 nodes with $m = 2$ (left) and for a GED graph of 1000 nodes with $\alpha = 1.5$ (right)

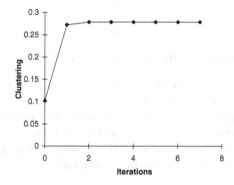

Fig. 9. Increase in clustering for a PLRG generated graph of 1000 nodes with $\alpha = 1.5$

figures, there is a marked increase in clustering for each example graph in few iterations.

We next determine the validity and accuracy of using an approximate value for the clustering of a graph instead of requiring that the exact value be know. There is obviously some error between the approximate and actual values, but according to sampling theory, increasing the sample size s will reduce this error. However, when the s exceeds a certain threshold, further increasing it does not significantly decrease the error. To determine a proper sample size, we provide the following lemma.

Lemma 1. *When the sample size $s = Z_{\frac{p}{2}}^2/(2E)^2$, the error between the approximate and actual clustering does not exceed E with probability at least $1 - p$.*

Proof. Denote by σ^2 the variance of node clustering in the graph, γ_s the sampled clustering of s nodes in the graph, and γ_a the actual clustering of the graph. According to sampling theory [24], when the sample size is:

$$s = \frac{Z_{\frac{\rho}{2}}^2 \sigma^2}{E^2} \ , \tag{9}$$

error $|\gamma_s - \gamma_a|$ does not exceed *error margin* E with probability $1 - \rho$, where $\rho \in (0, 1)$ is a *significance level* and $z_{\rho/2}$ is the positive z value that is at the vertical boundary for the area of $\rho/2$ in the right tail of the standard normal distribution.

Note that the clustering of each node is between 0 and 1. When half of the total nodes have clustering 0 and the other half have clustering 1, the variance of node clustering reaches its maximum point, where the clustering of the graph is 0.5. Therefore:

$$\sigma^2 \le \frac{\sum_{i=1}^n 0.5^2}{n} = 0.5^2 \ . \tag{10}$$

Using (9) and (10), we conclude that when sample size:

$$s = \frac{Z_{\frac{\rho}{2}}^2}{(2E)^2} \ , \tag{11}$$

$|\gamma_s - \gamma_a|$ does not exceed E with probability at least $1 - \rho$.

Thus by Lemma 1, we can determine the sample size s based on an error margin E and significance level ρ. For the error margin $E = 0.1$ and $\rho = 0.05$, we only need to sample $s = 1.96^2/0.2^2 \approx 97$ nodes to contain the absolute error within 0.1 of the correct value with probability at least 0.95. This result shows that we are indeed able to approximate the clustering of a graph with very few samples, which justifies its inclusion in our algorithm.

5   Conclusion

In this paper, we offer an algorithmic solution to the clustering problem and show that we can frequently improve this metric in graphs produced by existing degree-based generators to values well over 0.5. This in turn allows those generators to better simulate the AS-level Internet topology.

References

1. R. Albert and A. Barabasi, "Topology of Evolving Network: Local Events and Universality," *Physica Review Letters* 85, 2000.
2. R. Albert, H. Jeong, and A. Barabasi, "Diameter of the World Wide Web", *Nature* 401, 1999.
3. R. Albert, H. Jeong, and A. Barabasi, "Error and Attack Tolerance in Complex Networks," *Nature* 406, July 2000.
4. A. Barabasi, H. Jeong, R. Ravasz, Z. Neda, T. Vicsek, and A. Schubert, "On the Topology of the Scientific Collaboration Networks," *Physica A 311*, 2002.
5. W. Aiello, F. R. K. Chung, and L. Lu, "A Random Graph Model for Massive Graphs," *ACM STOC*, 2000.

6. W. Aiello, F. R. K. Chung, and L. Lu, "Random Evolution in Massive Graphs," *IEEE FOCS*, 2001.
7. A. Barabasi, R.Albert, and H.Jeong, "Mean-field Theory for Scale-free Random Networks," *Physica A* 272, 1999.
8. A. Barabasi and R. Albert, "Emergence of Scaling in Random Networks," *Science*, October 1999.
9. A. Barabasi, R. Albert, and Hawoong Jeong, "Scale-free Characteristics of Random Networks: The Topology of the World Wide Web," *Physica A 281*, 69-77 (2000).
10. T. Bu and D. Towsley, "On Distinguishing between Internet Power Law Topology Generators," *IEEE INFOCOM*, June 2002.
11. K. Calvert, M. Doar, and E. Zegura, "Modeling Internet Topology," *IEEE Communications Magazine*, June 1997.
12. H. Chang, R.Govindan, S. Jamin, S. Shenker, and W. Willinger, "Towards Capturing Representative AS-Level Internet Topologies," *University of Michigan Technical Report CSE-TR-454-02*, 2002.
13. Q. Chen, H. Chang, R. Govindan, S. Jamin, S. Shenker, and W. Willinger, "The Origin of Power-laws in Internet Topologies Revisited," *IEEE INFOCOM*, June 2002.
14. F. R. K. Chung, "Connected Components in Random Graphs with Given Expected Degree Sequences," *Annals of Combinatorics* 6, 2002.
15. F. R. K. Chung, "The Spectra of Random Graphs with Given Expected Degree," *http://math.ucsd.edu/ fan/wp/specp.pdf*.
16. M. Doar, "A Better Model for Generating Test networks," *IEEE GLOBECOM*, November 1996.
17. P. Erdos and A. Renyi, "On Random Graphs," *I, Publication Math.* Debrecen 6, 290-291, 1959.
18. T. S. E. Ng and H. Zhang, "Predicting Internet Network Distance with Coordinates-Based Approaches," *IEEE INFOCOM*, 2002.
19. M. Faloutsos, P. Faloutsos, and C. Faloutsos, "On power-Law Relationships of the Internet Topology," *ACM SIGCOMM*, August 1999.
20. P. Francis *et al.*, "IDMaps: A Global Internet Host Distance Estimation Service," *IEEE/ACM Transactions on Networking*, vol. 9, no. 5, October 2001.
21. C. Jin, Q. Chen and S. Jamin, "Inet: Internet Topology Generator," *University of Michigan Technical Report CSE-RT-433-00*, 2000.
22. G. Huston, "Architectural Requirements for Inter-Domain routing in the Internet," *IETF Draft draft-iab-bgparch-01.txt*.
23. C. Labovitz, A. Ahuja, R. Wattenhofer, and S. Venkatachary, "The Impact of Internet Policy and Topology on Delayed Routing Convergence," *IEEE INFOCOM*, 2001.
24. Y. Leon, "Probability and Statistics with applications," *International Textbook Company*, 1969.
25. http://mathworld.wolfram.com/HypergeometricFunction.html
26. M. Mihail and C.H. Papadimitriou, "On the Eigenvalue Power Law," *RANDOM*, 2002.
27. M. Mihail and N. Visnoi, "On Generating Graphs with Prescribed Degree Sequences for Complex Network Modeling applications," *ARACNE*, 2002.
28. M. Molloy and B. Reed, "A Critical Point for Random Graphs with a Given Degree Sequence," *Random Structures and Algorithms*, 6:161-180, 1995.
29. National Laboratory for Applied Network Research "Global ISP Interconnectivity by AS Number," *http://moat.nlanr.net/as/*.

30. P. Radoslavov, H. Tangmunarunkit, H. Yu, R. Govindan, S. Shenker, and D. Estrin, "On Characterizing Network Topologies and Analyzing Their Impact on Protocol Design," *USC Technical Report 00-731*, February 2000.
31. E. Ravasz and A. Barabasi, "Hierarchical Organization in Complex Networks," *Physical Review* E(in press).
32. D. J. Watts, "Small World," *Princeton University Press*, 1999.
33. S. Yook, H. Jeong, and A. Barabasi, "Modeling the Internet's Large-scale Topology," *Proceedings of the Nat'l Academy of Sciences* 99, 2002.

Generating Web Graphs with Embedded Communities

Vivek B. Tawde[1], Tim Oates[2], and Eric Glover[3]

[1] Yahoo! Research Labs, Pasadena, CA, USA
[2] University of Maryland Baltimore County, Baltimore, MD, USA
[3] Ask Jeeves, Emeryville, CA, USA

Abstract. We propose a model to generate a web graph with embedded communities. The web graph will also have the properties that are observed on the web like bow-tie structure, node reachability and link distributions.

We use a community as a basic building block to model the link distributions. We interconnect these communities using a model for interconnection that uses a citation matrix. The generated graph consisting of a combination of interlinked communities, is converted into a larger web graph by modifying the link distribution to achieve the desired distribution of the whole web.

Using communities as the basic building block helps in maintaining local properties like resiliency and reachability of the simulated web graph.

1 Introduction

The dynamics of the World-Wide-Web (WWW) make it interesting to study but hard to characterize. The evolution of the WWW is a complex combination of sociological behaviors which result in the formation of interesting substructures of web pages such as bipartite cores, web rings [15] and nepotistic connectivity [2, 7]. It also gives rise to communities in the web graph which are characterized by the link structure of the graph. The microscopic and macroscopic structural information of the web can be exploited while designing crawling and searching algorithms, and in understanding the phenomenon of the evolution of communities and their interaction.

Some of the characteristics of the WWW have been studied with respect to the web as a whole (e.g. diameter, connected components and bow-tie structure [4]) while some are given with respect to the substructure of the graph (e.g. link distributions across nodes [17], interconnection across substructures [6], interconnection of substructures to the rest of the graph).

1.1 Problem Definition

The ideal model for the web graph should encompass local as well as global characteristics of the web. Such a model will not only be useful in studying global struc-

S. Leonardi (Ed.): WAW 2004, LNCS 3243, pp. 80–91, 2004.

ture of the web but also in characterizing behavior of local components. It can be utilized to study the global effects of evolution of local components and vice-versa. Such a comprehensive model of the web graph will make it possible to reason about changes in properties of the graph in local or global contexts. It will also be a useful model to the research community which extensively studies algorithms for ranking web pages based on connectivity (e.g. HITS [14], Pagerank [3]).

In this work we attempt to build a model for the web that will capture not only its global structure but also maintain the properties of the local substructures in the web.

1.2   Related Work

One of the models for social networks proposed by Barabàsi and Albert [1] (the BA model) is based on *preferential attachment* that models the "rich get richer" behavior. Pennock et al. [17] improved this model by taking into consideration the *random attachment* factor along with preferential attachment which modeled the distribution of high as well as low connectivity nodes accurately for the entire web graph as well as for a subset of web pages. But it failed to consider interaction among groups of pages or unification of the groups to form larger groups.

Kumar et al. [15] studied the frequently occurring substructures in the web graph, like bipartite graphs, and found them to be topically similar to a great extent. This observation strengthens the argument that evolution of the web cannot be modeled via random graph generation, because a randomly generated graph has a very low probability of having structures like strongly connected components and bipartite cores which occur frequently in the web graph.

Broder et al. [4] revealed the "bow-tie" structure in the web graph which has a strongly connected component (SCC), 2 subgraphs connecting to and from SCC and isolated subgraphs. Dill et al. [9] discovered the bow-tie like structure in certain subgraphs on the web.

Flake et al. [11] have defined a community as a set of pages having more linkage within themselves than outside. The community discovery algorithm is based on the "max flow-min cut" theorem given by Ford and Fulkerson [12, 8].

The interaction among communities is studied by Chakrabarti et al. [6]. They model the topical interaction in the form of a $n \times n$ topic citation matrix for n topics where element $[i][j]$ is the probability that there is a link from a page on topic i to page on topic j.

1.3   Our Contribution

Our aim is to generate a web graph by using a combination of some of the global and local models described in section 1.2 that maintains global and local structure of the web.

We take a three step approach in generating the web graph:

1. Generate individual communities observed on the web using a model for preferential and random link distribution [17].

2. Combine the communities generated in the previous step using a model for interaction among communities(topic citation matrix [6]) to generate "combined community graph" (CCG).

3. Finally we convert the CCG into the web graph by achieving the link distribution of the entire web. This is done by adding nodes ("the rest of the web" (ROW)) with certain connectivity.

The choice of the modeling parameters and the modeling approach itself are based on existing models and empirical observations.

The proposed web graph model achieves the following goals:

1. Maintains link distributions of individual communities.

2. Combines individual communities to maintain their interaction level with respect to one another.

3. Generates a web graph from the "combined community graph" which will maintain observed global properties of the web.

2   Basic Building Blocks of the Web Graph Model

We use three different models in three different stages to produce the web graph: Model for Link Distribution of a Community; Model for Interaction among Communities and Model for Link Distribution of the entire Web.

2.1   Model for Link Distribution in a Community

Pennock et al. [17] have studied the link distribution of several communities on the web and found them to have a "unimodal" shape with a power-law tail which exhibits a "rich get richer" behavior (exhibited by the BA model [1]. Many social and biological processes do exhibit power-law behavior [5, 13, 16, 18]). Even though a power-law tail indicates preference for a small number of high connectivity nodes, the modal curve indicates that the low connectivity nodes get a relatively larger share of the links. Figure 1 shows the inlink distribution of *company homepages* which clearly depicts the unimodal link distribution curve with a power-law tail.

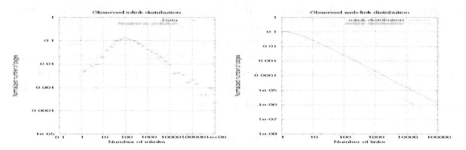

Fig. 1. Inlink Distribution of company homepages and Link Distribution of the Entire Web

Pennock et al. [17] propose a generative model to simulate the link distribution observed in communities based on *preferential attachment* and *random attachment*. The community begins with a certain number of nodes m_0. At every step one node and 2m edges are added to it (2m = average number of links per node). Model by Pennock et al. [17] choose the node to which the link will be assigned based on a mixture probability of *preferential attachment* and *random attachment* with mixture parameter α ($0 \leq \alpha \leq 1$). The expression for the probability that node i will be chosen to connect an edge is

$$\Pi(k_i) = \alpha \cdot \frac{k_i}{2mt} + (1 - \alpha) \cdot \frac{1}{m_0 + t} \tag{1}$$

where k_i = current connectivity of node i, and t = number of nodes added untill now. The parameters α and $2m$ are estimated from observed data.

Based on the above model, Pennock et al. [17] derived a closed form expression for the expected connectivity distribution for the generated graph using a "continuous mean-field approximation". It evaluates the probability of a node having connectivity k in a graph generated using equation 1. The expression is given below:

$$Pr(k) = [2m \cdot (1 - \alpha)]^{\frac{1}{\alpha}} \cdot [\alpha k + 2m \cdot (1 - \alpha)]^{-1 - \frac{1}{\alpha}} \tag{2}$$

To visualize the plot for $Pr(k)$ on a log-log scale, the expression is modified to display probability mass at each k. The expression is given below:

$$Pr(k) = \frac{ln10}{5} \cdot [2m \cdot (1 - \alpha)]^{\frac{1}{\alpha}} \cdot k \cdot [\alpha k + 2m \cdot (1 - \alpha)]^{-1 - \frac{1}{\alpha}} \tag{3}$$

Figure 1 shows example of observed link distribution across nodes and the analytical solution using equation 3.

Pennock et al. extended this model further to choose the source of a link according to α_{out} and choose the destination of that link according to α_{in}.

2.2   Model for Interaction Among Communities

Chakrabarti et al. [6] studied the relation between communities and proposed a model for interaction among communities. They modeled interaction among n communities using n X n matrix where element $[i][j]$ of the matrix gives the empirical probability that there is a link from a page in community i to a page in community j. Figure 2 from [6] shows the plot produced by Chakrabarti et al. [6] which shows the extent of interaction among 191 topics studied.

In our case, the estimation of element $[i][j]$ is based on the observed connections from pages in community i to pages in community j. This model can be used to interlink the various communities formed using any model (the BA model or Pennock et al. model) of link distribution.

2.3   Model for Link Distribution of the Web

We propose to use the Link Distribution of the Web as a benchmark for generating a web graph. This is a special case of the Link Distribution model proposed

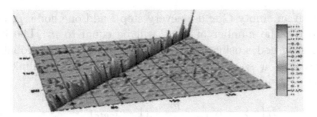

Fig. 2. Plot for 191 Topic Citation Matrix, reprinted from Chakrabarti et al. [6]

by Pennock et al. [17]. Pennock et al. studied the link distribution of the Web and estimated parameters for the entire web ($\alpha_{in} = 0.909$ and $\alpha_{in} = 0.581$). Broder et al. estimated the average inlink/outlink per page ($2m$) to be 8. Figure 1 shows the analytical curve produced using equation 3 with the parameters mentioned above.

3   Web Graph Model with Embedded Communities

In this section we explain the three step procedure to produce a web graph with communities embedded in it. We also explain how to map the two link distributions (inlink and outlink) produced in order to produce a web graph with bow-tie structure as observed on the existing web [4].

Expected Inputs: link distribution parameters for the communities; parameters for the interaction among communities (the citation matrix "CM"); parameters for the expected web graph
Expected Output: Simulated graph with embedded communities

3.1   Symbols and Notation

This is the list of symbols and notation that will be used in the following sections.
C - Community
$|M|$ - Number of member communities to be embedded in the graph
$m_0/m_0 + t$ - Initial/Existing number of nodes in C
p_i/q_i- Outlink/Inlink connectivity of node i
n_c - Number of nodes in community C
$p_c/q_c/m_c$- Average number of outlinks/inlinks/links per node of C
$\alpha_{outc}/\alpha_{inc}$ - Preferential attachment parameter for outlink/inlink distribution
p_w/q_w - Average number of outlinks/inlinks per node of the web

3.2   Modeling a Web Community

In this step, our goal is to produce a link distribution (outlink and inlink) for an n node community using α_{outc} and α_{inc} and to connect outlinks to inlinks within the community.

Generating Link Distribution: The procedure for generating link distributions of C is very similar to the one explained by Pennock et al. [17]. The idea is to start with an empty C and at every step t add one node, p_c outlinks and q_c inlinks to C. The number of steps will be equal to n_c. The links are assigned to existing nodes using Equation 1. For a node i, the probability that it will be chosen as a source/destination of some outlink/inlink is given by Equation 4/Equation 5.

$$\Pi(p_i) \; = \; \alpha_{outc} \cdot \frac{p_i}{p_c t} + (1 - \alpha_{outc}) \cdot \frac{1}{m_0 + t} \tag{4}$$

$$\Pi(q_i) \; = \; \alpha_{inc} \cdot \frac{q_i}{q_c t} + (1 - \alpha_{inc}) \cdot \frac{1}{m_0 + t} \tag{5}$$

The outlinks and inlinks are generated and distributed across n_c nodes independently. But none of the outlinks have been assigned destinations yet and none of the inlinks have been assigned sources yet. At this point C can be perceived as a collection of two sets of nodes, each of size n_c, with one of the sets acting as sources for outlinks and the other acting as destinations for inlinks. All the outlinks and inlinks are dangling (with one end not connected to any node).

To make a community graph out of these two sets, we need to map the elements of these two sets using a one-to-one mapping and also connect the dangling links.

Connecting Links within a Community: The citation matrix for the group of communities gives information about the extent of interlinking within the community. For C_i, the element $[i][i]$ of the citation matrix gives the fraction of the total outlinks of C_i pointing to nodes within C_i. So that fraction of the total dangling outlinks are selected at random and are connected to the randomly selected dangling inlinks of C_i.

Still at this point, we have C in the form of two sets of nodes, each of size n_c with one set having outlinks and the other with inlinks. But now due to interconnections made within C, not all outlinks and inlinks are dangling (Figure 3).

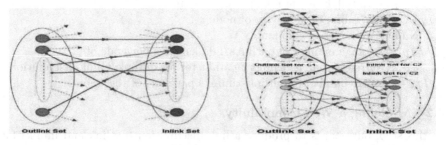

Fig. 3. A community generated according to a specific link distribution and interlinked within itself. And result of merging of communities C1 and C2

3.3   Merging Individual Communities

To model the interaction among all the $|M|$ communities $(C_1, C_2, ..., C_{|M|})$, we use the citation matrix CM which is an $|M|$ X $|M|$ matrix. Element $CM[i][j]$ is the fraction of total outlinks of C_i pointing to the nodes in C_j.

For every C_i, the outlinks in C_i are connected to the inlinks of C_j (for $j = 1...|M|, j \neq i$), using citation matrix CM. The element $CM[i][j] = f$ is the fraction of the total outlinks of C_i to be connected to the nodes in C_j. So f fraction of the total outlinks of C_i are selected at random and are connected to the randomly selected dangling inlinks of C_j. Figure 3 depicts the interconnection between two communities, which can be generalized to visualize the merging of $|M|$ communities.

All the $|M|$ communities merged and interconnected in the way described above form the CCG which will be used to build the web graph with embedded communities.

3.4   Converting a Combined Community Graph (CCG) to the Web Graph

To convert the CCG to the web graph, we need to alter its link distribution to match the benchmark web graph distribution and map nodes from outlink and inlink set.

Changing Link Distribution of the CCG: Nodes of certain connectivity are added to the CCG such that the link distribution of the modified CCG will match the benchmark link distribution. Equation 3 can be used to project the target size of the web graph based on current CCG and target benchmark distribution. And then number of additional nodes and their connectivity can be calculated. This analysis becomes simple since Equation 3 treats the range of connectivity as buckets of connectivity. This whole process can be visualised as modifing shape of a curve to match another shape by addition of data points in a certain way. These additional nodes form the "rest of the web" (ROW). (The details of the steps are omited due to space constraint.)

Linking the CCG to the Rest of the Web (ROW): Any dangling links in the CCG are randomly attached to nodes in ROW. The remaining dangling links in ROW are connected at random to each other.

Theoritically, the final graph created should have the total number of outlinks equal to the total number of inlinks. Our experimental results show that, in most cases, they are almost the same. The marginal difference is because of the random assignment of links (outlinks or inlinks) to the additional nodes that form ROW.

3.5   One-to-One Mapping Between Outlink and Inlink Nodes

The one-to-one mapping between outlink and inlink set results in nodes with both inlinks and outlinks. The way the mapping is done affects the properties of the resultant web graph like the bow-tie structure [4]. We use heurisitcs based

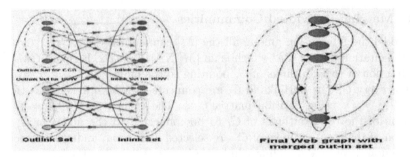

Fig. 4. Merging CCG and ROW. And Webgraph after one-to-one mapping

mapping since no other mapping (generative model by Pennock et al. or random mapping) could come close to replicating the observed bow-tie model.

We map the inlink nodes to outlink nodes in a way that will give us the closest possible replication of bow-tie and sizes of its various components. The heurisitcs of this mapping are based on the observed sizes of the bow-tie components and the link distribution of the web graph generated. Figure 4 shows the final state of the web graph after the mapping of nodes is complete.

4   Experiments and Results

We tested our model by generating a web graph using communities and compared it with the actual web graph. We experimented with fourteen different communities obtained from Yahoo! [19] categories.

For each community, the member pages were collected from the web and based on the observed data, the parameters for the link distributions (α_{outc}, α_{inc}, p_c and q_c) were estimated. The citation matrix for the fourteen communities was also generated from the observed community data. The parameters for the expected web distribution were taken from previous studies of the web (Pennock et al. [17] and Broder et al. [4])($\alpha_{outc} = 0.581$, $\alpha_{inc} = 0.909$and $p_c = q_c = 8$).

Link Distribution of Simulated Web Graph: We produced a simulated web graph by using various combinations of the fourteen communities. In each case, the simulated graph closely followed the link distribution of the real web.

Embedded Communities and the Size of the Simulated Web Graph: Table 1 indicate that the relative size of a community to the web graph (% size of Comm) is controlled by the α parameters and the average link parameters p_c and q_c. The closer they are to those for the web, the greater is the relative size of the community to the web. We found the average links (p_w and q_w) to be almost equal to 8 (Table 1) which is the benchmark for the target web distribution set. This further authenticates the accuracy of the model.

Bow-Tie Structure in the Simulated Web Graph: Broder et al. [4] measured size of various components of the bow-tie. We also studied the bow-tie

Table 1. Result of converting a Community/CCG to a Web Graph

Community-Name	$S(\times 10^6)$	p_w	q_w	% size of Comm	$(p_c + q_c)$
univ	3.3	8.5	9.3	0.04	2197
photographers	0.02	8	6.8	13	22
photographers+univ	0.12	8.3	7.7	3	850
All Combined	2.6	8.5	9	3	375

structure of the simulated web graph and determine the size of each component. Table 2 shows the comparison between the actual web and our model for the web.

Table 2. Bow-tie Model Comparison

	SCC	IN	OUT	TENDRIL	DISCONN	WCC
Real Web	27	20	20	20	13	87
Our Model N=1000	20	30	15	19	16	84
Our Model N=10000	21	32	15	16	16	84
Our Model N=50000	22	32	15	17	15	85

Our model came close to replicating the bow-tie on the real web. Some variation in the size of the SCC and OUT are most likely the result of the discrepency between observations made by Pennock et al. [17] and Broder et al. [4].

Table 3 shows that only heuristics based mapping came close to replicating the bow-tie structure of the real web.

Table 3. Mapping Schemes Comparison

Mapping	SCC	IN	OUT	TENDRIL	DISCONN	WCC
Real Web	27	20	20	20	13	87
Heuristic	21	30	15	16	16	84
Pennock et al.	32	43	2	0	23	77
Random(Seed=1000)	11	46	8	21	14	86

Power-law and Zipf Distribution of the Inlinks: We present the plots in Figure 5 for inlink distribution plotted as power-law distribution and also as Zipf distribution for the real web and the simulated web graph (N=100000). Both the plots show a notable deviation in the power-law in-degree distribution from the actual power-law reference as compared to Zipf in-degree distribution.

Resiliency of the Web Graph: Broder et al. [4] observed that around 90% of the web is weakly connected. To examine the strength of this wide spread connectivity, they removed all nodes with inlink counts greater than a certain k and then checked the size of the Weakly Connected Component (WCC). They observed that even though all nodes with k connectivity were removed, the size of

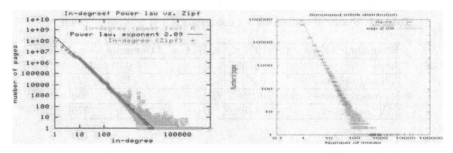

Fig. 5. Inlink Distribution for the Real Web (reprinted from Broder et al. [4]) and that for the Simulated Web Graph

Table 4. Change in Size of WCC on removal of k in-connectivity nodes

k	1000	100	10	5
Real Web	95%	90%	55%	30%
Our Model(N=10000)	90%	80%	55%	20%
Our Model(N=50000)	90%	81%	53%	23%

the WCC has only reduced marginally. The simulated web graph also exhibited similar resiliency (Table 4).

Reachability analysis of the Simulated Web Graph: Broder et al. [4] did the reachability analysis of the real web by starting a BFS (using outlinks, inlinks and both) on randomly selected nodes from the graph and then for each, observing the fraction of nodes reached. They plotted cumulative distributions of nodes covered in BFS (Figure 6 from [4]).

The performance of the reachability analysis on the simulated web graph is also very similar to the one observed on the real web (Figure 6, 4rth plot, N=100000, 500 starting nodes). The outlink and undirected reachability curves closely match those for the real web. About 60% of the nodes (SCC+IN) have high outlink reachability. About 83% of the nodes (SCC+IN+OUT) have high undirected reachability.

For about 60% of the nodes there is no external inlink connectivity. This is because, 60% of the nodes in the generated graph (using parameters from Pennock et al. [17]) do not have any inlinks (as stated in section 3.5). For the remaining 40% nodes (SCC+OUT) the reachability is high.

5   Conclusion and Future Work

To simulate a web graph, we used community structure as a basic building block. We modeled communities using existing models based on link distributions. We interconnected these communities using a model for interconnection which uses a citation matrix. Then the "combined community graph" is converted to a web

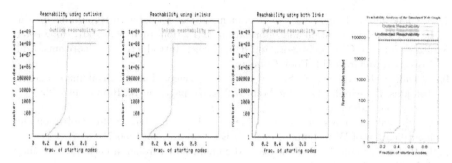

Fig. 6. Reachability Analysis of the Real Web Graph (reprinted from Broder et al. [4]) and the Simulated Web Graph

graph by modifying the link distribution to achieve the link distribution of the web. We used heuristics to impart the bow-tie structure to the simulated web graph.

Using community as a basic building block for a web graph helps in maintaining local properties of the web graph. The global link distribution of the simulated graph followed the observed link distribution. The simulated web graph also showed the resemblance to the actual web in terms of bow-tie structure, power-law and Zipf distributions, reachability of nodes and resiliency properties.

Future Work: It will be of tremendous value if instead of heuristics, a model for imparting bow-tie structure to the simulated graph can be found.

Using more communities to form larger fraction of the web graph will give a better idea about the scalability and accuracy of this model. It will be interesting to see if this model can be modified to considered overlapping communities.

It will be interesting to apply some of the existing ranking and resource discovery algorithms (for example Pagerank, HITS) and compare the results with the results on the actual web graph.

References

1. A.-L. Barabsi and R. Albert. Emergence of scaling in random networks. Science, 286:509-512, 1999.
2. K. Bharat and M. Henzinger. Improved Algorithms for topic distillation in a hyperlinked environment. In 21st International ACM SIGIR Conference on Research and Development in Information Retrieval, pages 104-111, Aug. 1998.
3. S. Brin and L. Page. The anatomy of a large-scale hypertextual web search engine. In Proceedings of the 7th World-Wide Web Conference (WWW7), 1998.
4. A. Broder, R. Kumar, F. Maghoul, P. Raghavan, S. Rajagopalan, R. Stata, A. Tomkins, and J. Wiener. Graph structure in the Web: Experiments and models. In WWW9, pages 309-320, Amsterdam, May 2000. Elsevier.
5. J. L. Casti Complexity 1, 12-15 (1995).
6. S. Chakrabarti, M. Joshi, K. Punera, and D. M. Pennock. The structure of broad topics on the Web. In WWW 2002, Hawaii.

7. S. Chakrabarti, Integrating the Document Object Model with Hyperlinks for Enhanced Topic Distillation and Information Extraction. In WWW 2001, Hong kong.
8. T. H. Cormen, C. E. Leiserson, and R. L. Rivest. Introduction to algorithms. MIT Press and McGraw-Hill Book Company.
9. S. Dill, S. R. Kumar, K. S. McCurley, S. Rajagopalan, D. Sivakumar, and A. Tomkins. Self-similarity in the Web. In VLDB, pages 69-78, Rome, Sept. 2001.
10. http:://www.dmoz.com/
11. G. W. Flake, S. Lawrence, C. Lee Giles, and F. M. Coetzee. Self-organization and identification of Web communities. IEEE Computer, 35(3):66-71, 2002.
12. L. R. Ford Jr. and D. R. Fulkerson. Maximum flow through a network. Canadian J. Math., 8:399-404, 1956
13. B. A. Huberman and L. A. Adamic (1999) Nature (London) 401, 131
14. J. Kleinberg. Authoritative sources in a hyperlinked environment. In Proc. ACM-SIAM Symposium on Discrete Algorithms, 1998. Also appears as IBM Research Report RJ 10076(91892).
15. S. R. Kumar, P. Raghavan, S. Rajagopalan, and A. Tomkins. Trawling the web for emerging cyber-communities. WWW8 / Computer Networks, 31(11-16):1481-1493, 1999.
16. R. M. May, (1988) Science 214, 1441-1449.
17. D. M. Pennock, G. W. Flake, S. Lawrence, C. L. Giles, and E. J. Glover. Winners don't take all: Characterizing the competition for links on the Web. Proceedings of the National Academy of Sciences, 2002.
18. S. Wasserman and K. Faust (1998) Social Network Analysis: Methods and Applications (Cambridge Univ. Press, Cambridge, U.K.)
19. http:://www.yahoo.com/
20. G.K. Zipf. Human Behavior and the Principle of Least Effort, Addison-Wesley, 1949.

Making Eigenvector-Based Reputation Systems Robust to Collusion

Hui Zhang[1], Ashish Goel[2], Ramesh Govindan[1],
Kahn Mason[2], and Benjamin Van Roy[2]

[1] University of Southern California, Los Angeles CA 90007, USA
{huizhang, ramesh}@usc.edu
[2] Stanford University, Stanford CA 94305, USA
{ashishg, kmason, bvr}@stanford.edu

Abstract. Eigenvector based methods in general, and Google's PageRank algorithm for rating web pages in particular, have become an important component of information retrieval on the Web. In this paper, we study the efficacy of, and countermeasures for, *collusions* designed to improve user rating in such systems.

We define a metric, called the amplification factor, which captures the amount of PageRank-inflation obtained by a group due to collusions. We prove that the amplification factor can be at most $1/\epsilon$, where ϵ is the reset probability of the PageRank random walk. We show that colluding nodes (e.g., web-pages) can achieve this amplification and increase their rank significantly in realistic settings; further, several natural schemes to address this problem are demonstrably inadequate.

We propose a relatively simple modification to PageRank which renders the algorithm insensitive to such collusion attempts. Our scheme is based on the observation that nodes which cheat do so by "stalling" the random walk in a small portion of the web graph and, hence, their PageRank must be especially sensitive to the reset probability ϵ. We perform exhaustive simulations on the Web graph to demonstrate that our scheme successfully prevents colluding nodes from improving their rank, yielding an algorithm that is robust to gaming.

1 Introduction

Reputation systems are becoming an increasingly important component of information retrieval on the Web. Such systems are now ubiquitous in electronic commerce, and enable users to judge the reputation and trustworthiness of online merchants or auctioneers. In the near future, they may help counteract the free-rider phenomenon in peer-to-peer networks by rating users of these networks and thereby inducing social pressure to offer their resources for file-sharing [6, 10, 12]. Also, they may soon provide context for political opinion in the Web logging (blogging) world, enabling readers to calibrate the reliability of news and opinion sources.

A simple, and common way to measure a user's reputation is to use a referential link structure, a graph where nodes represent entities (users, merchants,

S. Leonardi (Ed.): WAW 2004, LNCS 3243, pp. 92–104, 2004.
© Springer-Verlag Berlin Heidelberg 2004

authors of blogs) and links represent endorsements of one user by another. A starting point for an algorithm to compute user reputations might then be the class of eigenvector- or stationary distribution- based reputation schemes exemplified by the PageRank algorithm[1].

Algorithms based on link structure are susceptible to collusions; we make the notion of collusion more precise later, but for now we loosely define it as a manipulation of the link structure by a group of users with the intent of improving the rating of one or more users in the group. The PageRank algorithm published in the literature has a simple "resetting" mechanism which alleviates the impact of collusions. The PageRank value assigned to a page can be modeled as the fraction of time spent at that page by a random walk over the link structure; to reduce the impact of collusions (in particular, rank "sinks"), the algorithm resets the random walk at each step with probability ϵ.

In this paper, we define a quantity called the *amplification factor* that characterizes the amount of PageRank-inflation obtained by a group of colluding users. We show that nodes may increase their PageRank values by at most an *amplification factor* $\frac{1}{\epsilon}$; intuitively, a colluding group can "stall" the random walk for that duration before it resets. While this may not seem like much (a typical value for ϵ is 0.15), it turns out that the distribution of PageRank values is such that even this amplification is sufficient to significantly boost the *rank* of a node based on its PageRank value. What's worse is that all users in a colluding group could and *usually* do benefit from the collusion, so there is significant incentive for users to collude. For example, we found that it was easy to modify the link structure of the Web by having a low (say 10,000-th) ranked user collude[2] with a user of even lower rank to catapult themselves into the top-400. Similar results exist for links in other rank levels.

Two natural candidate solutions to this problem present themselves – identifying groups of colluding nodes, and identifying individual colluders by using detailed return time statistics from the PageRank random walk. The former is computationally intractable since the underlying optimization problems are NP-Hard. The latter does not solve the problem since we can identify scenarios where the return time statistics for the colluding nodes are nearly indistinguishable from those for an "honest" node.

How then, can PageRank based reputation systems protect themselves from such collusions? Observe that the ratings of colluding nodes are far more sensitive to ϵ than those of non-colluding nodes. This is because the PageRank values of colluding nodes are amplified by "stalling" the random walk; as explained before, the amount of time a group can stall the random walk is roughly $1/\epsilon$. This suggests a simple modification to the PageRank algorithm (called the *adaptive-*

---

[1] Although not viewed as such, PageRank may be thought of as a way of rating the "reputation" of web sites.

[2] Collusion implies intent, and our schemes are not able to determine intent, of course. Some of the collusion structures are simple enough that they can occur quite by accident.

resetting scheme) that allows different nodes to have different values of the reset probability. We have not been able to formally prove the correctness of our scheme (and that's not surprising given the hardness result), but we show, using extensive simulations on a real-world link structure, that our scheme significantly reduces the benefit that users obtain from collusion in the Web. Furthermore, while there is substantial intuition behind our detection scheme, we do not have as good an understanding of the optimum policy for modifying the individual reset probabilities. We defer an exploration of this to future work.

While we focus on PageRank in our exposition, we believe that our scheme is also applicable to other eigenvector-based reputation systems (e.g. [10, 12]). We should point out that the actual page ranking algorithms used by modern search engines (*e.g.*, Google) have evolved significantly and incorporates other domain specific techniques to detect collusions that are not (and will not be, for some time to come) in the public domain. But we believe that it is still important to study "open-source style" ranking mechanisms where the algorithm for rank computation is known to all the users of the system. Along with web-search, such an algorithm would also be useful for emerging public infrastructures (peer-to-peer systems and the blogosphere) whose reputation systems design are likely to be based on work in the public domain.

The remainder of this paper is organized as follows. We discuss related work in Section 2. In Section 3 we study the impact of collusions on the PageRank algorithm, in the context of the Web. Section 4 shows the hardness of making PageRank robust to collusions. In Section 5 we describe the *adaptive-resetting* scheme, and demonstrate its efficiency through exhaustive simulations on the Web graph. Section 6 presents our conclusions.

2   Related Work

Reputation systems have been studied heavily in non-collusive settings such as eBay [5, 14] – such systems are not the subject of study in this paper.

In the literature, there are at least two well-known eigenvector-based link analysis algorithms: HITS [11] and PageRank [16]. HITS was originally proposed to refine search outputs from Web search engines and discover the most influential web pages defined by the principal eigenvector of its link matrix. As discovering the principal eigenvector is the goal, original HITS doesn't assign a total ordering on the input pages, and collusion is less of a problem for it than for PageRank. On the contrary, PageRank was proposed to rank order input pages and handling clique-like subgraphs is a fundamental design issue.

Despite their difference, both algorithms have been applied into the design of reputation systems for distributed systems [10, 12]. These designs have mainly focused on the decentralization part, while their collusion-proofness still relies on the algorithm itself.

In the context of topic distillation on the Web, many extensions to PageRank and HITS algorithms [2, 4, 8, 9, 13] have been proposed for improving search-query results. Two general techniques - content analysis, and bias ranking with a seed link set - are used to handle problematic (spam) and irrelevant web links. While working well in their problem space, these approaches do not give answers to the algorithmic identification of collusions in a general link structure.

Ng *et al.* [15] studied the stability of HITS and PageRank algorithm with the following question in mind: when a small portion of the given graph is removed (*e.g.*, due to incomplete crawling), how severely do the ranks of the remaining pages change, especially for those top ranked nodes? They show that HITS is sensitive to small perturbations, while PageRank is much more stable. They proposed to incorporate the PageRank's "reset-to-uniform-distribution" into HITS to enhance its stability.

Finally, for context, we briefly describe the original PageRank algorithm with its random walk model. Given a directed graph, a random walk W starts its journey on any node with the same probability. At the current node x, with probability $(1 - \epsilon)$ W jumps to one of the nodes that have links from x (the choice of neighbor is uniform), and with probability ϵ, W decides to restart (*reset*) its journey and again choose any node in the graph with the same probability. Asymptotically, the stationary probability that W is on node x is called the PageRank value of x, and all nodes are ordered based on the PageRank values.

In the rest of the paper, we use the term *weight* to denote the PageRank (PR) value, and *rank* to denote the ordering. We use the convention that the node with the largest PR weight is ranked first.

3   Impact of Collusions on PageRank

In this section, we first show how a *group* of nodes could modify the referential link structure used by the PageRank algorithm in order to boost their PageRank weights by up to $1/\epsilon$. We then demonstrate that it is possible to induce simple collusions in real link structures (such as that in the Web) in a manner that raises the *ranking* of colluding nodes[3] significantly.

3.1   Amplifying PageRank Weights

In what follows, we will consider the PageRank algorithm as applied to a directed graph $G = (V, E)$. $N = |V|$ is the number of the nodes in G. A node in G corresponds, for example, to a Web page in the Web graph, or a blog in the blog graph; an edge in G corresponds to a reference from one web page to another, or from one blog to another. Let $d(i)$ be the out-degree of node i, and $W_v(i)$ be the *weight* that the PageRank algorithm computes for node i. We define on each edge $e_{ij} \in E$ the weight $W_e(e_{ij}) = \frac{W_v(i) \times (1-\epsilon)}{d(i)}$.

Let $V' \subset V$ be a set of nodes in the graph, and let G' be the subgraph induced by V'. E' is defined to be the set of all edges e_{ij} such that at least one of i and j is in V'. We classify the edges in E' into three groups:

---

[3] We use "pages", "nodes" and "users" interchangeably in the rest of the paper.

In Links: An edge e_{ij} is an in link for G' if $i \notin V'$ and $j \in V'$. E'_{in} denotes the set of in links of G'.

Internal Links: An edge e_{ij} is an internal link for G' if $i \in V'$ and $j \in V'$. $E'_{internal}$ denotes the set of internal links of G'.

Out Links: An edge e_{ij} is an out link for G' if $i \in V'$ and $j \notin V'$. E'_{out} denotes the set of out links of G'.

One can then define two types of weights on G':

- $W_{in}(G') = \sum_{e:e \in E'_{in}} W_e(e) + \frac{N'}{N}$, $N = |V|$, $N' = |V'|$.
- $W_G(G') = \sum_{v:v \in V'} W_v(v)$.

Intuitively, $W_{in}(G')$ is, in some sense, the "actual" weight that should be assigned to G', when G' is regarded in its entirety (*i.e.* as one unit). On the other hand, $W_G(G')$ is the total "reputation" of the group that would be assigned by PageRank. Note that nodes within G' can boost this reputation by manipulating the link structure of the internal links or the out links.

Then, we can define a metric we call the *amplification factor* of a graph G as $Amp(G) = \frac{W_G(G)}{W_{in}(G)}$. Given this definition, we prove (see Appendix A in the companion technical report [17] for the proof) the following theorem:

Theorem 1. *In the original PageRank system,* $\forall G' \subseteq G, Amp(G') < \frac{1}{\epsilon}$.

3.2   PageRank Experiments on the Real-World Graph

It might not be surprising to find out that the weight inflation in PageRank groups could be as high as $\frac{1}{\epsilon}$, since it's already known from [15] that eigenvector-based reputation systems are not stable under link structure perturbation. However, it's not clear what is the practical import of amplifying PageRank weights. Specifically, is it easy for a group of colluding nodes to achieve the upper bound of the amplification factor, $\frac{1}{\epsilon}$? Can nodes improve their *ranking* significantly?

To answer these questions, we obtained a large Web subgraph from the Stanford WebBase [18]. It contains upwards of 80 million URLs, and is called $\mathcal{W}$ in the rest of the paper. We then modified one or more subgraphs in $\mathcal{W}$ to simulate collusions, and measured the resulting PageRank weights for each node. We tried a few different modifications, and report the results for one such experiment.

Our first experiment on $\mathcal{W}$ is called *Collusion200*. This models a small number of web pages *simultaneously* colluding. Each collusion consists of a pair of nodes with *adjacent* ranks. Such a choice is more meaningful than one between a low ranked node and a high ranked node, since the latter could have little incentive to collude. Each pair of nodes removes their original out links and adds one new out link to each other. In the experiment reported here, we induce 100 such collusions at nodes originally ranked around 1000th, 2000th, ..., 100000th.

There is a subtlety in picking these nodes. We are given a real-world graph in which there might already be colluding groups (intentional or otherwise). For this reason, we carefully choose our nodes such that they are unlikely to be already colluding (the precise methodology for doing this will become clear in Section 5.2 when we describe how we can detect colluding groups in graphs).

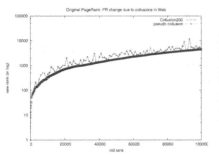

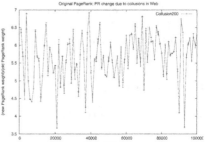

Fig. 1. $\mathcal{W}$: New PR rank after *Collusion200*

Fig. 2. $\mathcal{W}$: New PR weight (normalized by old PR weight) after *Collusion200*

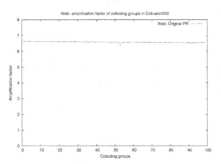

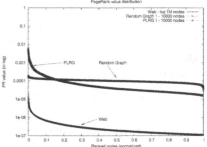

Fig. 3. Amplification factors of the 100 colluding groups in Collusion200

Fig. 4. PR distribution in 3 topologies

We calculate the PageRank weights and ranks for all nodes before (called old rank and weight) and after (called new rank and weight) *Collusion200* on $\mathcal{W}$ with $\epsilon = 0.15$ (a default value assumed in [16]). Figures 1 & 2 show the rank and weight change for those colluding nodes. In addition, we also plot in Figure 1 the rank that each colluding node could have achieved if its weight were amplified by $\frac{1}{\epsilon}$ while all other nodes remained unchanged in weight, which we call *pseudo collusion*.

As we can see, all colluding nodes increased their PR weight by at least 3.5 times, while the majority have a weight amplification over 5.5. More importantly, collusion boosts their ranks to be more than 10 times higher and close to the best achievable. For example, a colluding node originally ranked at 10002th had a new rank at 451th, while the 100005th node boosted its rank to 5033th by colluding with the 100009th node, which also boosted its rank to 5038th.

Thus, even concurrent, simple (2-node) collusions of nodes with comparable original ranks can result in significant rank inflation for the colluding nodes. For another view of this phenomenon in Figure 3 we plot the amplification factors achieved by the colluding groups in $\mathcal{W}$. It clearly shows that almost all colluding groups attain the upper bound.

But what underlies the significant *rank inflation* in our results? Figure 4 shows the PageRank weight distribution of $\mathcal{W}$ (only top 1 million nodes for interest). It also includes, for calibration, the PageRank weight distribution on power low random graph (PLRG) [1] and the classical random graph [3] topologies. First, observe that the random graph has a flat curve, which implies that in such a topology, almost any nodes could take one of the top few positions by amplifying its weight by $\frac{1}{\epsilon}$. Secondly, $\mathcal{W}$ and *PLRG* share the same distribution characteristic, *i.e.*, the top nodes have large weights, but the distribution flattens quickly after that. This implies that in these topologies, low ranked nodes can inflate their ranks by collusion significantly (though perhaps not to the top 10).

While we have discussed only one experiment with a simple collusion scheme, there are many other schemes through which nodes can successfully achieve large rank inflation (Section 5.2 presents a few such schemes). We believe, however, that our finding is both general (*i.e.*, not constrained to the particular types of collusions investigated here) and has significant practical import since $\mathcal{W}$ represents a non-trivial portion of the Web. Having established that collusions can be a real problem, we now examine approaches to making the PageRank algorithm robust to collusions.

4   On the Hardness of Making PageRank Robust to Collusions

We will now explore two natural approaches to detecting colluding nodes, and demonstrate that neither of them can be effective.

The first approach is to use finer statistics of the PageRank random walk. Let the random variable X_v denote the number of time steps from one visit of node v to the next. It is easy to see that the PageRank value of v is exactly $1/\mathbf{E}[X_v]$ where $\mathbf{E}[X_v]$ denotes the expectation of X_v. For the simplest collusion, where two nodes A and B delete all their out-links and start pointing only to each other, the random walk will consist of a long alternating sequence of A's and B's, followed by a long sojourn in the remaining graph, followed again by a long alternating sequence of A's and B's, and so on [4] . Clearly, X_A is going to be 2 most of the time, and very large (with high probability) occasionally. Thus, the ratio of the variance and the expectation of X_A will be disproportionately large. It is now tempting to suggest using this ratio as an indicator of collusion.

Unfortunately, there exist simple examples (such as large cycles) where this approach fails to detect colluding nodes. We will present a more involved example where not just the means and the variances, but the *entire distributions* of X_H and X_C are nearly identical; here H is an "honest" node and C is cheating to improve its PageRank. The initial graph is a simple star topology. Node 0 points to each of the nodes $1 \ldots N$ and each of these nodes points back to node 0 in turn. Now, node N starts to cheat; it starts colluding with a new node $N + 1$

---

[4] Incidentally, it is easy to show that this collusion mode can achieve the theoretical upper bound of $1/\epsilon$ on the amplification factor.

so that N and $N + 1$ now only point to each other. The new distributions X_0 and X_N can be explicitly computed, but the calculation is tedious. Rather than reproduce the calculation, we provide simulation results for a specific case, where $N = 7$ and $\epsilon = 0.12$. Figure 5 shows the revisit distribution for nodes 0 (the original hub) and 7 (the cheating node). The distributions are nearly identical. Hence, any approach that relies solely on the detailed statistics of X_v is unlikely to succeed.

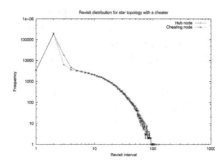

Fig. 5. Frequency of revisit intervals for the cheating node (node 7) and the honest node (node 0) for the star-topology. The simulation was done over 1,000,000 steps

Thus, a more complete look at the graph structure is needed, one that factors in the various paths the random walk can take. One natural approach to identifying colluders would be to directly find the subgraph with the maximum amplification (since colluders are those with high amplification). However, it is very unlikely that this problem is tractable. Consider the intimately related problem of finding a group S of size k which maximizes the difference of the weights, $W_G(S) - W_{in}(S)$, rather than the ratio. This problem is NP-Hard via reduction to the densest k-subgraph problem [7]. Details of the reduction are in the companion technical report [17]. There are no good approximation algorithms known for the densest k-subgraph problem (the best known is $O(N^{1/3})$). The reduction is approximation preserving. Hence, identifying colluding groups is unlikely to be computationally tractable even in approximate settings.

This suggests that our goals should be more modest – rather than identifying the entire colluding group, we focus on finding individual nodes that are cheating. This is the approach we take in the next section.

5 Heuristics for Making PageRank Robust to Collusions

Given our discussion of the hardness of making PageRank robust to collusions, we now turn our attention to heuristics for achieving this. Our heuristic is based on an observation explained using the following example. Consider a small (compared to the size of the original graph) group S of colluding nodes. These nodes can not influence links from $V - S$ into S. Hence, the only way these nodes can increase their stationary weight in the PageRank random walk is by stalling the random walk *i.e.* by not letting the random walk escape the group. But in the PageRank algorithm, the random walk resets at each node with probability ϵ. Hence, colluding nodes must suffer a significant drop in PageRank as ϵ increases.

This forms the basis for our heuristic for detecting colluding nodes. We expect the stationary weight of colluding nodes to be highly correlated [5] with $1/\epsilon$ and that of non-colluding nodes to be relatively insensitive to changes in ϵ . While our hypothesis can be analytically verified for some interesting special cases (details in the companion technical report [17]), we restrict ourselves to experimental evidence in this paper.

5.1   The Adaptive-Resetting Heuristic

The central idea behind our heuristic for a collusion-proof PageRank algorithm is that the value of the reset probability is *adapted*, for each node, to the degree of collusion that the node is perceived to be engaged in. This *adaptive-resetting* scheme consists of two phases:

1. Collusion detection
   (a) Given the topology, calculate the PR weight vector under different ϵ values.
   (b) Calculate the correlation coefficient between the curve of each nodes x's PR weight and the curve of $\frac{1}{\epsilon}$. Label it as $co\text{-}co(x)$, which is our proxy for the collusion of x. $co\text{-}co(x) = co\text{-}co(x) < 0 \ ? \ 0 : \ co\text{-}co(x)$.
2. ϵ Personalization
   (a) Now the node x's *out-link* personalized-$\epsilon = F(\epsilon_{default}, co - co(x))$.
   (b) The PageRank algorithm is repeated with these personalized-ϵ values.

The function $F(\epsilon_{default}, co - co(x))$ provides a knob for a system designer to appropriately punish colluding nodes. In our experiments we tested two functions:

Exp. function $F_{Exp} = \epsilon_{default}^{(1.0-\text{CO-CO}(x))}$.
Linear function $F_{Linear} = \epsilon_{default} + (0.5 - \epsilon_{default}) \times \text{co-co}(x)$.

The choice of function is subjective and application-dependent, and given space limitations, we mostly present results based on F_{Exp}.

5.2   Experiments

As in Section 3, we conducted experiments on the $\mathcal{W}$ graph. In all experiments with our adaptive-resetting scheme, we chose seven ϵ values in the *collusion*

---

[5]   The correlation coefficient of a set of observations $(x_i, y_i) : i = 1, .., n$ is given by

$$\text{co-co}(x, y) = \frac{\sum_{i=1,\cdots,n}(x_i - \overline{x})(y_i - \overline{y})}{\sqrt{\sum_{i=1,\cdots,n}(x_i - \overline{x})^2 \sum_{i=1,\cdots,n}(y_i - \overline{y})^2}}.$$

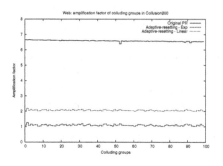
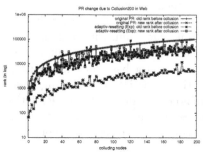

Fig. 6. $\mathcal{W}$: amplification factors of the 100 colluding groups in Collusion200

Fig. 7. $\mathcal{W}$: PR rank comparison between original PageRank and Adaptive-ϵ scheme in *Collusion200*

detection phase – $0.6, 0.45, 0.3, 0.15, 0.075, 0.05$, and 0.0375 – and used 0.15 as $\epsilon_{default}$. While there are eight PageRank calculations, the actual computational time for the adaptive-resetting scheme was only 2-3 times that of the original PageRank algorithm. This is because the computed PR weight vector for one ϵ value is a good initial state for the next ϵ value.

Basic Experiment: We first repeated the experiment *Collusion200* for adaptive-resetting scheme. As mentioned in Section 3.2, all the colluding nodes are chosen from the nodes unlikely to be already colluding, and this is judged by their *co-co* values in the original topology. Precisely, we select nodes with $co - co(x) \leq 0.1$. Choosing nodes with arbitrary *co-co* values doesn't invalidate the conclusions in this paper (as discusses in the companion technical report [17]), but our selection methodology simplifies the exposition of our scheme.

We compared the original PageRank algorithm, the adaptive-resetting schemes using F_{Exp} and F_{Linear}. As shown in Figure 6, the adaptive-resetting scheme F_{Exp} restricted the amplification factors of the colluding groups to be very close to one, and F_{Linear} also did quite well compared to the original PageRank.

In Figure 7 we compare the original PageRank and the adaptive-resetting scheme using F_{Exp} based on the old and new rank before and after *Collusion200* in $\mathcal{W}$. For the original PageRank algorithm the rank distribution clearly indicates how nodes benefit significantly from collusion. The curves for the adaptive-resetting scheme nearly overlap, illustrating the robustness of our heuristic. Furthermore, note that the curves of the PageRank algorithm before collusions and the adaptive-resetting before collusions are close to each other, which means the weight of non-colluding nodes is not affected noticeably when applying the adaptive-resetting scheme instead of the original PageRank scheme.

Other Collusion Topologies an Experiment with Miscellaneous Collusion Topologies: We tested adaptive-resetting scheme under other collusion topologies in an experiment called *Collusion22*. In *Collusion22* 22 low co-co (≤ 0.1) nodes are selected for 3 colluding groups:

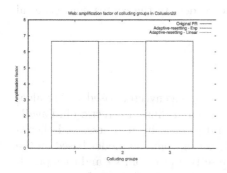
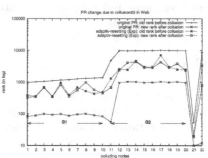

Fig. 8. $\mathcal{W}$: Amplification factors of the 3 colluding groups in Collusion22

Fig. 9. $\mathcal{W}$: PR rank comparison between original PageRank and Adaptive-ϵ scheme in *Collusion22*

G1. $G1$ has 10 nodes, which remove their old out links and organize into a single-link ring. All nodes have their original ranks at around 1000th.

G2. $G2$ has 10 nodes, which remove their old out links and organize into a star topology by one hub pointing to the other 9 nodes and vice versa. The hub node has its original rank at around 5000th, while the other nodes are ranked at around 10000th originally.

G3. $G3$ has 2 nodes, which remove the old out links and organize into a circle. One is originally ranked at around 50th, and the other at around 9000th.

We ran *Collusion22* on $\mathcal{W}$ using both original PageRank and adaptive-resetting scheme. We first observed that the adaptive-resetting scheme *successfully detected all 22 colluding nodes* by reporting high *co-co* values (> 0.96).

In Figure 8, we compare the original PageRank algorithm, the adaptive-resetting schemes with function F_{Exp} and F_{Linear} based on the metric *amplification factor* under *Collusion22*. As in Figure 6 the two adaptive-resetting schemes successfully restricted the weight amplification for the colluding nodes.

In Figure 9 we compare original PageRank and adaptive-resetting scheme with function F_{Exp} based on the old and new rank before and after *Collusion22* in $\mathcal{W}$. The results for the graph $\mathcal{B}$ were similar and are omitted. As we can see, the nodes of $G1$ and $G2$ seem to have some rank improvement in adaptive-resetting before collusions compared to their ranks in the PageRank algorithm before collusion, while their weights have increased only marginally. This is due to the rank drop of many high rank nodes with high *co-co* values in *adaptive-resetting before collusions*. Lastly, it is interesting to observe that with the original PageRank algorithm, even the two nodes with significantly different ranks in $G3$ can benefit mutually from a simple collusion: the 8697th node rocketed to the 12th, and the 54th node also jumped to the 10th position.

More Experiment Results: We have done more experiments to validate the correctness of adaptive-resetting scheme. Due to the space limit, we do not present them here and refer the interested readers to the companion technical report [17] for details.

6   Conclusion

In this paper we studied the robustness of one eigenvector-based rating algorithm: PageRank. We point out the importance of collusion detection in PageRank based reputation systems for real-world graphs, its hardness, and then a heuristic solution. Our solution involves detecting colluding nodes based on the sensitivity of their PageRank value to the resetting probability ϵ and then penalizing them by assigning them a higher reset probability. We have done extensive simulations on the Web graph to demonstrate the efficacy of our heuristic.

Acknowledgement

We would like to thank the Stanford Webbase group for making a pre-processed copy of the Web link structure available to us.

References

1. W. Aiello, F. Chung, and L. Lu. *A Random Graph Model for Massive Graphs.* the 32nd Annual Symposium on Theory of Computing, 2000.
2. K. Bharat, M. R. Henzinger. *Improved Algorithms for Topic Distillation in a Hyperlinked Environment.* Proceedings of SIGIR98, 1998.
3. B. Bollobas. *Random Graphs.* Academic Press, Inc. Orlando, Florida, 1985.
4. S. Chakrabarti, B. Dom, P. Raghavan, S. Rajagopalan, D. Gibson, J. Kleinberg. *Automatic resource compilation by analyzing hyperlink structure and associated text.* Proceeedings of the Seventh International Conference on World Wide Web,, 1998.
5. C. Dellarocas. *Analyzing the economic efficiency of eBay-like online reputation reporting mechanisms.* In Proceedings of the 3rd ACM Conference on Electronic Commerce, pages 171–179, 2001.
6. D. Dutta, A. Goel, R. Govindan, H. Zhang. *The Design of A Distributed Rating Scheme for Peer-to-peer Systems.* First Workshop on Economics of Peer-to-peer Systems, UC Berkeley, June 2003.
7. U. Feige and M. Seltser, *On the densest k-subgraph problems.* Technical report no. CS97-16, Department of Applied Math and Comp. Sci., Weizmann Institute, 1997.
8. Z. Gyongyi, H. Garcia-Molina, and J. Pedersen. *Combating Web Spam with TrustRank.* Technical Report, Stanford University, 2004.
9. T. Haveliwala. *Topic-sensitive PageRank.* In Proceedings of WWW2002, 2002.
10. S. Kamvar, M. Schlosser and H. Garcia-Molina. *EigenRep: Reputation Management in P2P Networks.* in Proc. WWW Conference, 2003.
11. J. M. Kleinberg. *Authoritative Sources in a Hyperlinked Environment.* Journal of the ACM, Vol.46, No.5, P604-632, 1999.
12. H.T. Kung, C.H. Wu. *Differentiated Admission for Peer-to-Peer Systems: Incentivizing Peers to Contribute Their Resources.* First Workshop on Economics of Peer-to-peer Systems, UC Berkeley, June 2003.

13. L. Li , Y. Shang , W. Zhang. *Improvement of HITS-based algorithms on web documents.* Proceedings of WWW2002, 2002.
14. N. Miller, P. Resnick, and R. Zeckhauser. *Eliciting honest feedback in electronic markets.* RWP02-39, Harvard Kennedy School of Government, 2002.
15. A. Y. Ng, A. X. Zheng, and M. I. Jordan. *Link analysis, eigenvectors, and stability.* International Joint Conference on Artificial Intelligence (IJCAI), 2001.
16. L. Page, S. Brin, R. Motwani, and T. Winograd. *The PageRank Citation Ranking: Bringing Order to the Web.* Stanford Digital Library Technologies Project, 1998.
17. The companion technical report is available at *ftp://ftp.usc.edu/pub/csinfo/tech-reports/papers/04-817.pdf*
18. The Stanford WebBase Project. *http://www-diglib.stanford.edu/testbed/doc2/WebBase/*

Towards Scaling Fully Personalized PageRank[*]

Dániel Fogaras[1,2] and Balázs Rácz[1,2]

[1]Computer and Automation Research Institute of the
Hungarian Academy of Sciences
[2]Budapest University of Technology and Economics
fd@cs.bme.hu, bracz+p31@math.bme.hu

Abstract Personalized PageRank expresses backlink-based page quality around user-selected pages in a similar way as PageRank expresses
quality over the entire Web. Existing personalized PageRank algorithms
can however serve on-line queries only for a restricted choice of page
selection. In this paper we achieve full personalization by a novel algorithm that computes a compact database of simulated random walks;
this database can serve arbitrary personal choices of small subsets of
web pages. We prove that for a fixed error probability, the size of our
database is linear in the number of web pages. We justify our estimation
approach by asymptotic worst-case lower bounds; we show that exact
personalized PageRank values can only be obtained from a database of
quadratic size.

1 Introduction

The idea of topic sensitive or personalized ranking appears since the beginning of
the success story of Google's PageRank [5, 23] and other hyperlink-based centrality measures [20, 4]. Topic sensitivity is either achieved by precomputing modified
measures over the entire Web [13] or by ranking the neighborhood of pages containing the query word [20]. These methods however work only for restricted
cases or when the entire hyperlink structure fits into the main memory.

In this paper we address the computational issues of personalized PageRank
[13, 18]. Just as all hyperlink based ranking methods, PageRank is based on the
assumption that *the existence of a hyperlink $u \to v$ implies that page u votes for
the quality of v.* Personalized PageRank (PPR) [23] enters user preferences by
assigning more importance to edges in the neighborhood of certain pages at the
user's selection. Unfortunately the naive computation of PPR requires a power
iteration algorithm over the entire web graph, making the procedure infeasible
for an on-line query response service.

We introduce a novel scalable Monte Carlo algorithm for PPR that precom-
putes a compact database. As described in Section 2, the database contains
simulated random walks, and PPR is estimated on-line with a limited number

---

[*] Research was supported by grants OTKA T 42559 and T 42706 of the Hungarian
National Science Fund, and NKFP-2/0017/2002 project Data Riddle.

of database accesses. Earlier algorithms [14] restricted personalization to a few topics, a subset of popular pages or to hosts; our algorithm on the other hand enables personalization for *any* small set of pages. Query time is linear in the number of pages with non-zero personalization. Similar statement holds for the previous approaches, too.

The price that we pay for full personalization is that our algorithm is randomized and less precise; the formal analysis of the error probability is discussed in Section 3. In Section 4 it is verified that we have to pay the price of approximation by showing that full personalization requires a database of $\Omega(V^2)$ bits over a graph with V vertices.

Though this approximation approach might fail or need longer query time in certain cases (for example for pages with large neighborhoods), the available personalization algorithms can be combined to resolve these issues. For example we can precompute personalization vectors for topics (topic-sensitive PR), popular pages with large neighborhoods (use [18]), some often requested combination of popular pages (sub-topics), and use our algorithm for those many pages not covered so far. This combination gives adequate precision for most queries with large flexibility for personalization.

Related Results. The possibility of personalization was first mentioned in [5, 23] together with PageRank. Linearity of PPR [13] implies that if PPR is precomputed for some preference vectors, then PPR can be calculated on-line for any linear combination of the preference vectors by combining the precomputed PPRs. Thus personalization was achieved for any combination of 16 basic topics in [13]; an experimental implementation is already available at [12]. The methods of [18] precompute PPR for at most 100.000 individual pages, and then for any subset of the individual pages personalization is available by linearity. Furthermore, the algorithm of [19] personalizes PageRank over hosts rather than single web pages. Instead of user preferences, [25] tunes PageRank automatically using the query keywords.

To the best of our knowledge randomized algorithms are not very common in the link-mining community. A remarkable exception [24] applies probabilistic counting to estimate the neighborhood function of web pages. Besides link-mining the paper [8] estimates the size of transitive closure for massive graphs occurring in databases. For text-mining algorithms [6] estimates the resemblance and containment of documents with a sampling technique.

Random walks were used before to compute various web statistics, mostly focused on sampling the web (uniformly or according to static PR) [16, 26, 1, 15], but also for calculating page decay [2] and similarity values [11].

The lower bounds of Section 4 show that precise PPR requires significantly larger database than Monte Carlo estimation does. Analogous results with similar communication complexity arguments were proved in [17] for the space complexity of several data stream graph algorithms.

Preliminaries. In this section we briefly introduce notation, and recall definitions and basic facts about PageRank. Let $\mathcal{V}$ denote the set of web pages, and

$V = |\mathcal{V}|$ the number of pages. The directed graph with vertex set $\mathcal{V}$ and edges corresponding to the hyperlinks will be referred to as the *web graph*. Let A denote the adjacency matrix of the webgraph with normalized rows and $c \in (0,1)$ the *teleportation constant*. In addition, let r be the so called *preference vector* inducing a probability distribution over $\mathcal{V}$. *PageRank* vector p is defined as the solution of the following equation [23]

$$p = (1 - c) \cdot pA + c \cdot r .$$

If r is uniform over $\mathcal{V}$, then p is referred to as the *global PageRank vector*. For non-uniform r the solution p will be referred to as *personalized PageRank vector* denoted by $\text{PPV}(r)$. The special case when for some page u the uth coordinate of r is 1 and all other coordinates are 0, the PPV will be referred to as the *individual PageRank vector* of u denoted by $\text{PPV}(u)$. Furthermore the vth coordinate of $\text{PPV}(u)$ will be denoted by $\text{PPV}(u,v)$.

Theorem 1 (Linearity, [13]). *For any preference vectors r_1, r_2, and positive constants α_1, α_2 with $\alpha_1 + \alpha_2 = 1$ the following equality holds:*

$$\text{PPV}(\alpha_1 \cdot r_1 + \alpha_2 \cdot r_2) = \alpha_1 \cdot \text{PPV}(r_1) + \alpha_2 \cdot \text{PPV}(r_2).$$

Linearity is a fundamental tool for scalable on-line personalization, since if PPV is available for some preference vectors, then PPV can be easily computed for any combination of the preference vectors. Particularly, for full personalization it suffices to compute individual $\text{PPV}(u)$ for all $u \in \mathcal{V}$, and the individual PPVs can be combined on-line for any small subset of pages. Therefore in the rest of this paper we investigate algorithms to make all individual PPVs available on-line.

The last statement of the introduction will play a central role in our PPV estimations. The theorem provides an alternate probabilistic characterization of individual PageRank scores.[1]

Theorem 2 ([18, 10]). *Suppose that a number L is chosen at random with probability $\Pr\{L = i\} = c(1 - c)^i$ for $i = 0, 1, 2, \ldots$ Consider a random walk starting from some page u and taking L steps. Then*

$$\text{PPV}(u,v) = \Pr\{\text{the random walk ends at page } v \}$$

2   Personalized PageRank Algorithm

In this section we will present a new Monte-Carlo algorithm to compute approximate values of personalized PageRank utilizing the above probabilistic characterization of PPR.

Definition 1 (Fingerprint). *A fingerprint of a vertex u is a random walk starting from u; the length of the walk is of geometric distribution of parameter*

---

[1] Notice that this characterization slightly differs from the random surfer formulation [23] of PageRank.

c, i.e. after every step the walk ends with probability c, and takes a further step with probability $1 - c$.

By Theorem 2 the ending vertex of a fingerprint, as a random variable, has the distribution of the personalized PageRank vector of u. We will calculate N independent fingerprints by simulating N independent random walks starting from u and approximate $PPV(u)$ with the empirical distribution of the ending vertices of these random walks. The ending vertices of the fingerprints will constitute the *index database*, and the output ranking will be computed at query time from the fingerprints of positive personalization weights using the linearity theorem.

To increase the precision of the approximation of $PPV(u)$ we will use the fingerprints of u's neighbors in the calculation, as described in Section 2.3.

The challenging problem is how to scale the indexing, i.e. how to generate N independent random walks for each vertex of the web graph. We assume that the edge set can only be accessed as a data stream, sorted by the source page, and we will count the database scans and total I/O size as the efficiency measure of our algorithms. The assumption is made, since even with the latest compression techniques [3] it does not seem plausible to store the entire web graph in main memory. Under such assumption it is infeasible to generate the random walks one-by-one, as it would require random access to the edge-structure.

We will consider two computational environments here: a single computer with constant random access memory (external memory algorithm) and a distributed system with tens to thousands of medium capacity computers. Both algorithms use similar techniques to the respective I/O efficient algorithms computing PageRank [7].

As the task is to generate N independent fingerprints, the single computer solution can be trivially parallelized to make use of a large cluster of machines, too. (Commercial web search engines have up to thousands of machines at their disposal.) Also, the distributed algorithm can be emulated on a single machine, which may be more efficient due to the different approach.

2.1   External Memory Indexing

We will incrementally generate the entire set of random walks simultaneously. Assume that the first k vertices of all the random walks (of length at least k) are already generated. At any time it is enough to store the starting and the current vertices of the fingerprint, as we are interested in adding the ending vertex to the index of the starting vertex. Sort these pairs by the ending vertices. Then by simultaneously scanning through the edge set and this sorted set we can have access to the neighborhoods of the current ending vertices, thus we can generate the random out-neighbor (the next vertex) of each partial fingerprint. For each partial fingerprint we also toss a biased coin to determine if it has reached its final length (with probability c) or has to advance to the next round (with probability $1 - c$). This algorithm is formalized as Algorithm 1.

The number of I/O operations the external memory sorting takes is $D \log_M D$, where D is the database size and M is the available main memory. Thus the I/O requirement of the sorting parts can be upper bounded by

Algorithm 1 Indexing (external memory method)

V is the number of vertices in the web graph, N is the required number of finger-prints for each vertex. For a vertex v, OutEdges[v] is the set of links on page v. The teleportation constant of PPR is c.

> **for** $u := 1$ to V **do**
>   **for** $i := 1$ to N **do**
>   Paths[$(u-1) \cdot N + i$].PathStart:=u
>   Paths[$(u-1) \cdot N + i$].PathEnd:=u
>   Fingerprint[u]:=$\emptyset$
> **while** Paths $\neq \emptyset$ **do**
>   sort Paths by PathEnd /*use an external memory sort*/
>   **for** $j := 1$ to Paths.length **do** /*simultaneous scan of OutEdges and Paths*/
> l :=random(OutEdges[Paths[j].PathEnd].length) /*choose a random edge*/
>   Paths[j].PathEnd:=OutEdges[Paths[j].PathEnd][l] /*prolong the path*/
>   **if** random()$< c$ **then** /*this fingerprint ends here*/
>   Fingerprint[Paths[j].PathStart].push(Paths[j].PathEnd)
>   Paths.delete(j)

$$\sum_{k=0}^{\infty}(1-c)^k NV \log_M((1-c)^k NV) = \frac{1}{c} NV \log_M(NV) - \Theta(NV)$$

using the fact that after k rounds the expected size of the Paths array is $(1 - c)^k NV$ (Recall that V and N denote the numbers of vertices and fingerprints respectively).

We need a sort on the whole index database to avoid random-access writes to the Fingerprint arrays. Also, upon updating the PathEnd variables we do not write the unsorted Paths array to disk, but pass it directly to the next sorting stage. Thus the total I/O is at most $\frac{1}{c} NV \log_M NV$ plus the necessary edge-scans.

Unfortunately this algorithm apparently requires as many edge-scans as the length of the longest fingerprint path, which can be very large: Pr{the longest fingerprint is shorter, than L} $= (1 - (1-c)^L)^{N \cdot V}$. Thus instead of scanning the edges in the final stages of the algorithm we will change strategy when the Paths array has become sufficiently small. Assume a partial fingerprint path has its current vertex at v. Then upon this condition the distribution of the end of this path is identical to the distribution of the end of any fingerprint of v. Thus to finish the partial fingerprint we can retrieve an already finished fingerprint of v. Although this decreases the number of available fingerprints for v, this results in only a very slight loss of precision.[2]

An other approach to this problem is to truncate the paths at a given length L and approximate the ending distribution with the static PageRank vector, as described in Section 2.3.

---

[2] Furthermore, we can be prepared for this event: the distribution of these v vertices will be close to the static PageRank vector, thus we can start with generating somewhat more fingerprints for the vertices with high PR values.

2.2 Distributed Index Computing

In the distributed computing model we will invert the previous approach, and instead of sorting the path ends to match the edge set we will partition the edge set of the graph in such a way that each participating computer can hold its part of the edges in main memory. So at any time if a partial fingerprint with current ending vertex v requires a random out-edge of v, it can ask the respective computer to generate one. This will require no disk access, only network transfer.

More precisely, each participating computer will have several queues holding (PathStart, PathEnd) pairs: one (large) input queue, and for each computer one small output queue[3].

The computation starts with each computer filling their own input queue with N copies of the initial partial fingerprints (v, v), for each vertex v belonging to the respective computer in the vertex partition.

Then in the input queue processing loop a participating computer takes the next input pair, generates a random out-edge from PathEnd, decides whether the fingerprint ends there, and if it does not, then places the pair in the output queue determined by the next vertex just generated.[4] If the output queue reaches the size of a network packet's size, then it is flushed and transferred to the input queue of the destination computer.

The total size of all the input and output queues equals the size of the Paths array in the previous approach after the respective number of iterations. The total network transfer can be upper bounded by $\sum_{n=0}^{\infty}(1 - c)^n NV = \frac{1}{c}NV$, if every fingerprint path needs to change computer in each step. As the web graph tends to have many 'local' links, with a suitable partition of vertices[5] the network transfer will be considerably less. Also note that this amount of transfer is distributed almost uniformly between all pairs of network nodes[6], so the effective switching capacity of the network is challenged, not the capacity of the individual network links.

2.3 Query

The basic query algorithm is as follows: to calculate $PPV(u)$ we load the ending vertices of the fingerprints for u from the index database, calculate the empirical distribution over the vertices, multiply it with $1 - c$, and add c weight to vertex u. This requires one database access (disk seek).

---

[3] Preferably the size of a network packet.

[4] Either we have to store the partition index for those v vertices that have edges pointing to in the current computer's graph, or part(v) has to be computable from v, for example by renumbering the vertices according to the partition.

[5] It should be enough to have each domain on a single computer, as the majority of the links are intra-domain links [19, 9].

[6] Also depending on the actual partition; as a heuristics one should use a partition that distributes the global PageRank uniformly across computers: the expected value of the total InQueue hits of a computer is proportional to the the total PageRank score of vertices belonging to that computer.

Algorithm 2 Indexing (distributed computing method)

The algorithm of one participating computer. Each computer is assumed to have a part of the OutEdges[] arrays, in memory. For a vertex v, part(v) is the index of the computer that has the out-edges of v. The queues hold pairs of vertices: (PathStart, PathEnd).

> **for** v s.t. part(v) = current computer **do**
>> InQueue.push((v, v))
>
> **while** at least one queue is not empty **do** /*some of the fingerprints are still being calculated*/
>> p :=InQueue.get()/*If empty, flush output queues and block until a packet arrives.*/
>> l :=random(OutEdges[p.PathEnd].length) /*choose an edge at random*/
>> q.PathEnd:=OutEdges[q.PathEnd][l] /*prolong the path*/
>> **if** random()$< c$ **then** /*teleport: this fingerprint ends here*/
>>> Fingerprint[q.PathStart].push(q.PathEnd)
>>
>> **else**
>>> o := part(q.PathEnd)
>>> OutQueue[o].push(p)
>>> **if** OutQueue[o].length $\geq$ reasonable packet size **then**
>>>> transmit OutQueue[o] to the InQueue of computer o.
>
> transmit the finished fingerprints to the proper computers for collecting and sorting.

To reach a precision beyond the number of fingerprints saved in the database we can use the recursive property of PPV [18]:

$$\text{PPV}(u) = c\mathbb{1}_u + (1 - c) \sum_{v \in O(u)} \text{PPV}(v)$$

where $\mathbb{1}_u$ denotes the measure concentrated at vertex u (i.e. the unit vector of u), and $O(u)$ is the set of out-neighbors of u.

This gives us the following algorithm: upon a query u we load the fingerprint endings for u, the set of out-edges of u, and the fingerprint endings for the vertices linked by u.[7] From this set of fingerprints we use the above equation to approximate $\text{PPV}(u)$ using a higher amount of samples, thus achieving higher precision. This is a tradeoff between query time (database accesses) and precision: with k database accesses we can approximate the vector from kN samples.

The increased precision is essential in approximating the PPV of a page with large neighborhood, as from N samples at most N pages will have positive approximated PPR values. Fortunately, this set is likely to contain the pages with highest PPR scores. Using the samples of the neighboring vertices will give more adequate result, as it will be formally analyzed in the next section.

We could also use the expander property of the web graph: after not so many random steps the distribution of the current vertex will be close to the static PageRank vector. Instead of allowing very long fingerprint paths we could com-

---

[7] We can iterate this recursion if we want to have even more samples.

bine the PR vector with coefficient $(1 - c)^L$ to the approximation and drop all fingerprints longer than L. This would also solve the problem of the approximated individual PPV vectors having many zeroes (in those vertices that have no fingerprints ending there). The indexing algorithms would benefit from this truncation, too.

There is a further interesting consequence of the recursive property. If it is known in advance that we want to personalize over a fixed (maybe large) set of pages, we can introduce an artificial node into the graph with the respective set of neighbors to generate fingerprints for that combination.

3 How Many Fingerprints Are Needed?

In this section we will discuss the convergence of our estimates, and analyze the required amount of fingerprints for proper precision.

It is clear by the law of large numbers that as the number of fingerprints $N \to \infty$, the estimate $\widehat{PPV}(u)$ converges to the actual personalized PageRank vector $PPV(u)$. To show that the rate of convergence is exponential, recall that each fingerprint of u ends at v with probability $PPV(u, v)$, where $PPV(u, v)$ denotes the vth coordinate of $PPV(u)$. Therefore $N \cdot \widehat{PPV}(u, v)$, the number of fingerprints of u that ends at v, has binomial distribution with parameters N and $PPV(u, v)$. Then Chernoff's inequality yields the following bound on the error of over-estimating $PPV(u, v)$ and the same bound holds for under-estimation:

$$\Pr\{\widehat{PPV}(u, v) > (1 + \delta)\, PPV(u, v)\} = \Pr\{N\, \widehat{PPV}(u, v) > N(1 + \delta)\, PPV(u, v)\}$$
$$\leq e^{-N \cdot PPV(u,v) \cdot \delta^2 / 4}.$$

Actually, for applications the exact values are not necessary. We only need that the ordering defined by the approximation match fairly closely the ordering defined by the personalized PageRank values. In this sense we have exponential convergence too:

Theorem 3. *For any vertices* u, v, w *consider* $PPV(u)$ *and assume that:*

$$PPV(u, v) > PPV(u, w)$$

Then the probability of interchanging v *and* w *in the approximate ranking tends to 0 exponentially in the number of fingerprints used.*

Theorem 4. *For any* $\epsilon, \delta > 0$ *there exists an* N_0 *such that for any* $N \geq N_0$ *number of fingerprints, for any graph and any vertices* u, v, w *such that* $PPV(u, v) - PPV(u, w) > \delta$, *the inequality* $\Pr\{\widehat{PPV}(u, v) < \widehat{PPV}(u, w)\} < \epsilon$ *holds.*

Proof. We prove both theorems together. Consider a fingerprint of u and let Z be the following random variable: $Z = 1$, if the fingerprint ends in v, $Z = -1$ if the fingerprint ends in w, and $Z = 0$ otherwise. Then $\mathbb{E}Z = \mathrm{PPV}(u,v) - \mathrm{PPV}(u,w) > 0$. Estimating the PPV values from N fingerprints the event of interchanging v and w in the rankings is equivalent to taking N independent Z_i variables and having $\sum_{i=1}^N Z_i < 0$. This can be upper bounded using Bernstein's inequality and the fact that $\mathrm{Var}(Z) = \mathrm{PPV}(u,v) + \mathrm{PPV}(u,w) - (\mathrm{PPV}(u,v) - \mathrm{PPV}(u,w))^2 \leq \mathrm{PPV}(u,v) + \mathrm{PPV}(u,w)$:

$$\Pr\{\tfrac{1}{N}\textstyle\sum_{i=1}^N Z_i < 0\} \leq e^{-N\frac{(\mathbb{E}Z)^2}{2\mathrm{Var}(Z)+4/3\mathbb{E}Z}}$$
$$\leq e^{-N\frac{(\mathrm{PPV}(u,v)-\mathrm{PPV}(u,w))^2}{10/3\,\mathrm{PPV}(u,v)+2/3\,\mathrm{PPV}(u,w)}}$$
$$\leq e^{-0.3N(\mathrm{PPV}(u,v)-\mathrm{PPV}(u,w))^2}$$

From the above inequality both theorems follow. □

The first theorem shows that even a modest amount of fingerprints are enough to make distinction between the high, medium and low ranked pages according to the personalized PageRank scores. However, the order of the low ranked pages will usually not follow the PPR closely. This is not surprising, and actually a deep problem of PageRank itself, as [22] showed that PageRank is unstable around the low ranked pages, in the sense that with little perturbation of the graph a very low ranked page can jump in the ranking order somewhere to the middle.

The second statement has an important theoretical consequence. When we investigate the asymptotic growth of database size as a function of the graph size, the number of fingerprints remains constant for fixed ϵ and δ.

4 Lower Bounds for PPR Database Size

In this section we will prove several lower bounds on the complexity of personalized PageRank. In particular, we will prove that exact computation the necessary index database size of a fully personalized PageRank must be at least $\Omega(V^2)$ bits, and if personalizing only for H pages, the database size is at least $\Omega(H \cdot V)$. In the approximate problem the lower bound for full personalization is linear in V, which is achieved by our algorithm of Section 2.

More precisely we will consider two-phase algorithms: in the first phase the algorithm has access to the edge set of the graph and has to compute an index database. In the second phase the algorithm gets a query of arbitrary vertices u, v (and w), and it has to answer based only on the index database. In this model we will lower bound the index database size. We will consider the following types of queries:

(1) Exact: Calculate $\mathrm{PPV}(u,v)$, the vth element of the personalized PageRank vector of u.
(2) Approximate: Estimate $\mathrm{PPV}(u,v)$ with a $\widehat{\mathrm{PPV}}(u,v)$ such, that for fixed $\epsilon, \delta > 0$
$$\Pr\{|\widehat{\mathrm{PPV}}(u,v) - \mathrm{PPV}(u,v)| < \delta\} \geq 1 - \epsilon$$
(3) Positivity: Decide whether $\mathrm{PPV}(u,v)$ is positive with error probability at most ϵ.

(4) Comparison: Decide in which order v and w are in the personalized rank of u with error probability at most ϵ.

(5) ϵ–δ comparison: For a fixed $\epsilon > 0, \delta > 0$ decide the comparison problem with error probability at most ϵ, if $|\text{PPV}(u, v) - \text{PPV}(u, w)| > \delta$ holds.

Our tool towards the lower bounds will be the asymmetric communication complexity game *bit-vector probing* [17]: there are two players A and B, A has an m-bit vector x, B has an index $y \in \{1, 2, \ldots, m\}$, and they have to compute the function $f(x, y) = x_y$, i.e. the output is the yth bit of the input vector. To compute the proper output they have to communicate, and communication is restricted in the direction $A \to B$. The *one-way communication complexity* [21] of this function is the required bits of transfer in the worst case for the best protocol.

Theorem 5 ([17]). *Any protocol that outputs the correct answer to the bit-vector probing problem with probability at least $\frac{1+\gamma}{2}$ must transmit at least γm bits.*

Now we are ready to prove our lower bounds. In all our theorems we assume that personalization is calculated for H vertices, and there are V vertices in total. In the case of full personalization $H = V$.

Theorem 6. *Any algorithm solving the positivity problem (3) must use an index of size $\Omega((1 - 2\epsilon)HV)$ bits.*

Proof. Set $\frac{1+\gamma}{2} = 1 - \epsilon$. We give a communication protocol for the bit-vector probing problem. Given an input bit-vector x we will create a graph, that 'codes' the bits of this vector. Player A will create a PPV index on this graph, and transmit this index to B. Then Player B will use the positivity query algorithm for some vertices (depending on the requested index y) such that the answer to the positivity query will be the yth bit of the input vector x. Thus if the algorithm solves the PPV indexing and positivity query with error probability ϵ, then this protocol solves the bit-vector probing problem with probability $\frac{1+\gamma}{2}$, so the transferred index database's size is at least γm.

For the $H \leq V/2$ case consider the following graph: let $u_1, \ldots, u_H$ denote the vertices for whose the personalization is calculated. Add $v_1, v_2, \ldots, v_n$ more vertices to the graph. Let the input vector's size be $m = H \cdot n$. In our graph each vertex v_j has a loop, and for each $1 \leq i \leq H$ and $1 \leq j \leq n$ the edge (u_i, v_j) is in the graph iff bit $(i - 1)n + j$ is set in the input vector.

For any index $1 \leq y \leq m$ let $y = (i-1)n+j$; the personalized PageRank value $\text{PPV}(u_i, v_j)$ is positive iff (u_i, v_j) edge was in the graph indexed, thus iff bit y was set in the input vector. If $H \leq V/2$ the theorem follows since $n = V - H = \Omega(V)$ holds implying that $m = H \cdot n = \Omega(H \cdot V)$ bits are 'coded'.

Otherwise, if $H > V/2$ the same construction proves the statement with setting $H = V/2$. $\qquad\square$

Corollary 1. *Any algorithm solving the exact PPV problem (1) must have an index database sized $\Omega(H \cdot V)$ bits.*

Theorem 7. *Any algorithm solving the approximation problem (2) needs an index database of $\Omega(\frac{1-2\epsilon}{\delta}H)$ bits on a graph with $V = H + \Omega(\frac{1}{\delta})$ vertices. For smaller δ the index database requires $\Omega((1 - 2\epsilon)HV)$ bits.*

Proof. We will modify the construction of Theorem 6 for the approximation problem. We have to achieve that when a bit is set in the input graph, then the queried $\mathrm{PPV}(u_i, v_j)$ value should be at least 2δ, so that the approximation will decide the positivity problem, too. If the u_i vertex in the input graph of our construction has k edges connected to it, then each of those v_j end-vertices will have exactly $\frac{1-c}{k}$ weight in $\mathrm{PPV}(u_i)$. For this to be over 2δ we can have at most $n = \frac{1-c}{2\delta}$ possible $v_1, \ldots, v_n$ vertices. With $\frac{1+\gamma}{2} = 1 - \epsilon$ the theorem follows.

For small δ the original construction suffices. $\square$

This radical drop in the storage complexity is not surprising, as our approximation algorithm achieves this bound (up to a logarithmic factor): for fixed ϵ, δ we can calculate the necessary number of fingerprints N, and then for each vertex in the personalization we store exactly N fingerprints, independently of the graph's size.

Theorem 8. *Any algorithm solving the comparison problem (4) requires an index database of $\Omega((1 - 2\epsilon)HV)$ bits.*

Proof. We will modify the graph of Theorem 6 so that the existence of the specific edge can be queried using the comparison problem. To achieve this we will introduce a third set of vertices in the graph construction $w_1, \ldots, w_n$, such that w_j is the complement of v_j: A puts the edge (u_i, w_j) in the graph iff (u_i, v_j) was not put into, which means bit $(i - 1)n + j$ was not set in the input vector.

Then upon query for bit $y = (i - 1)n + j$, consider $\mathrm{PPV}(u_i)$. In this vector exactly one of v_j, w_j will have positive weight (depending on the input bit x_y), thus the comparison query $\mathrm{PPV}(u_i, v_j) > \mathrm{PPV}(u_i, w_j)$ will yield the required output for the bit-vector probing problem. $\square$

Corollary 2. *Any algorithm solving the ϵ–δ comparison problem (5) needs an index database of $\Omega(\frac{1-2\epsilon}{\delta}H)$ bits on a graph with $V = H + \Omega(\frac{1}{\delta})$ vertices. For smaller δ the index database needs $\Omega((1 - 2\epsilon)HV)$ bits.*

Proof. Modifying the proof of Theorem 8 according to the proof of Theorem 7 yields the necessary results. $\square$

5   Conclusions

In this paper we introduced a new algorithm for calculating personalized Page-Rank scores. Our method is a randomized approximation algorithm based on simulated random walks on the web graph. It can provide full personalization with a linear space index, such that the error probability converges to 0 exponentially with increasing the index size. The index database can be computed even on the scale of the entire web, thus making the algorithms feasible for commercial web search engines.

We justified this relaxation of the personalized PageRank problem to approximate versions by proving quadratic lower bounds for the full personalization problems. For the estimated PPR problem our algorithm is space-optimal up to a logarithmic factor.

Acknowledgement

We wish to acknowledge András Benczúr, Katalin Friedl and Tamás Sarlós for several discussions and comments on this research. Furthermore we are grateful to Glen Jeh for encouraging us to consider Monte-Carlo PPR algorithms.

References

1. Z. Bar-Yossef, A. Berg, S. Chien, J. Fakcharoenphol, and D. Weitz. Approximating aggregate queries about web pages via random walks. In *Proceedings of the 26th International Conference on Very Large Data Bases*, pages 535-544. Morgan Kaufmann Publishers Inc., 2000.
2. Z. Bar-Yossef, A. Z. Broder, R. Kumar, and A. Tomkins. Sic transit gloria telae: towards an understanding of the web's decay. In *Proceedings of the 13th World Wide Web Conference (WWW)*, pages 328-337. ACM Press, 2004.
3. P. Boldi and S. Vigna. The webgraph framework I: compression techniques. In *Proceedings of the 13th World Wide Web Conference (WWW)*, pages 595-602. ACM Press, 2004.
4. A. Borodin, G. O. Roberts, J. S. Rosenthal, and P. Tsaparas. Finding authorities and hubs from link structures on the world wide web. In *10th International World Wide Web Conference*, pages 415-429, 2001.
5. S. Brin and L. Page. The anatomy of a large-scale hypertextual Web search engine. *Computer Networks and ISDN Systems*, 30(1-7):107-117, 1998.
6. A. Z. Broder. On the resemblance and containment of documents. In *Proceedings of the Compression and Complexity of Sequences (SEQUENCES'97)*, pages 21-29. IEEE Computer Society, 1997.
7. Y.-Y. Chen, Q. Gan, and T. Suel. I/O-efHcient techniques for computing PageRank. In *Proceedings of the eleventh international conference on Information and knowledge management*, pages 549-557. ACM Press, 2002.
8. E. Cohen. Size-estimation framework with applications to transitive closure and reachability. *J. Comput. Syst. Sci.*, 55(3):441-453, 1997.
9. N. Eiron and K. S. McCurley. Locality, hierarchy, and bidirectionality in the web. In *Second Workshop on Algorithms and Models for the Web-Graph (WAW 2003)*, 2003.
10. D. Fogaras. Where to start browsing the web? In *3rd International Workshop on Innovative Internet Community Systems I2CS, published as LNCS 2877/2003*, pages 65-79, 2003.

11. D. Fogaras and B. Rácz. A scalable randomized method to compute link-based similarity rank on the web graph. In *Proceedings of the Clustering Information over the Web workshop, Conference on Extending Database Technology*, 2004. http://www.ilab.sztaki.hu/websearch /Publications/index.html.

12. P.Google, http://labs.google.com/personalized.

13. T. H. Haveliwala. Topic-sensitive PageRank. In *Proceedings of the 11th World Wide Web Conference (WWW)*, Honolulu, Hawaii, 2002.

14. T. H. Haveliwala, S. Kamvar, and G. Jeh. An analytical comparison of approaches to personalizing PageRank. Technical report, Stanford University, 2003.

15. M. R. Henzinger, A. Heydon, M. Mitzenmacher, and M. Najork. Measuring index quality using random walks on the Web. In *Proceedings of the 8th World Wide Web Conference, Toronto, Canada*, pages 213-225, 1999.

16. M. R. Henzinger, A. Heydon, M. Mitzenmacher, and M. Najork. On near-uniform url sampling. In *Proceedings of the 9th international World Wide Web conference on Computer networks*, pages 295-308, 2000.

17. M. R. Henzinger, P. Raghavan, and S. Rajagopalan. Computing on data streams. *External memory algorithms*, pages 107-118, 1999.

18. G. Jeh and J. Widom. Scaling personalized web search. In *Proceedings of the 12th World Wide Web Conference (WWW)*, pages 271-279. ACM Press, 2003.

19. S. Kamvar, T. H. Haveliwala, C. Manning, and G. Golub. Exploiting the block structure of the web for computing PageRank. Technical report, Stanford University, 2003.

20. J. Kleinberg. Authoritative sources in a hyperlinked environment. *Journal of the ACM*, 46(5):604-632, 1999.

21. E. Kushilevitz and N. Nisan. *Communication complexity*. Cambridge University Press, 1997.

22. R. Lempel and S. Moran. Rank stability and rank similarity of link-based web ranking algorithms in authority connected graphs. In *Second Workshop on Algorithms and Models for the Web-Graph (WAW 2003)*, 2003.

23. L. Page, S. Brin, R. Motwani, and T. Winograd. The PageRank citation ranking: Bringing order to the web. Technical report, Stanford Digital Library Technologies Project, 1998.

24. C. R. Palmer, P. B. Gibbons, and C. Faloutsos. ANF: a fast and scalable tool for data mining in massive graphs. In *Eighth ACM SIGKDD International Conference on Knowledge Discovery and Data Mining*, pages 81-90. ACM Press, 2002.

25. M. Richardson and P. Domingos. The Intelligent Surfer: Probabilistic combination of link and content information in PageRank. *Advances in Neural Information Processing Systems*, 14:1441-1448, 2002.

26. P. Rusmevichientong, D. M. Pennock, S. Lawrence, and C. L. Giles. Methods for sampling pages uniformly from the world wide web. In *AAAI Fall Symposium on Using Uncertainty Within Computation*, pages 121-128, 2001.

Fast PageRank Computation Via a Sparse Linear System
(Extended Abstract)

Gianna M. Del Corso[1], Antonio Gullí[1,2,*], and Francesco Romani[1]

[1] Dipartimento di Informatica, University of Pisa, Italy
[2] IIT-CNR, Pisa

Keywords: Link Analysis, Search Engines, Web Matrix Reducibility and Permutation.

1 Introduction and Notation

The research community has devoted an increased attention to reduce the computation time needed by Web ranking algorithms. Many efforts have been devoted to improve PageRank [4, 23], the well known ranking algorithm used by Google. The core of PageRank exploits an iterative weight assignment of ranks to the Web pages, until a fixed point is reached. This fixed point turns out to be the (dominant) eigenpair of a matrix derived by the Web Graph itself. Brin and Page originally suggested to compute this pair using the well-known Power method [12] and they also gave a nice interpretation of PageRank in terms of Markov Chains. Recent studies about PageRank address at least two different needs. First, the desire to reduce the time spent weighting the nodes of the Web Graph which takes several days. Second, to assign *many* PageRank values to each Web page, as results of PageRank's personalization [14–16] that was recently presented by Google as beta-service (see http://labs.google.com/personalized/). The Web changes very rapidly and more than 25% of links are changed and 5% of "new content" is created in a week [8]. This result indicates that search engines need to update link based ranking metrics (such as PageRank) very often and that a week-old ranking may not reflect very well the current importance of the pages. This motivates the need of accelerating the PageRank computation. Previous approaches followed different directions such as the attempt to compress the Web Graph to fit it into main memory [3], or the implementation in external memory of the algorithms [13, 7]. A very interesting research track exploits efficient numerical methods to reduce the computation time. These kind of numerical techniques are the most promising and we have seen many intriguing results in the last few years to accelerate the Power iterations [18, 13, 21]. In the literature [1, 21, 23] are presented models which treat in a different way pages

---

* Work partially supported by the Italian Registry of ccTLD.it.

with no out-links. In this paper we consider the original PageRank model (see Section 2) and, by using numerical techniques, we show that this problem can be transformed in an equivalent linear system of equations, where the coefficient matrix is as sparse as the Web Graph itself. This new formulation of the problem makes it natural to investigate the structure of the sparse coefficient matrix in order to exploit its reducibility. Moreover, since many numerical iterative methods for linear system solution can benefit by a reordering of the coefficient matrix, we rearrange the matrix increasing the data locality and reducing the number of iterations needed by the solving methods (see Section 4). In particular, we evaluate the effect of many different permutations and we apply several methods such as Power, Jacobi, Gauss-Seidel and Reverse Gauss-Seidel [25], on each of the rearranged matrices. The disclosed structure of the permuted matrix, makes it possible to use block methods which turn out to be more powerful than the scalar ones. We tested our approaches on a Web Graph crawled from the net of about 24 million nodes and more than 100 million links. Our best result, achieved by a block method, is a reduction of 58% in Mflops and of 89% in time with the respect of the Power method taken as reference method to compute the PageRank vector.

We now give some notations and definitions that will be useful in the rest of the paper. Let M by an $n \times n$ matrix. A scalar λ and a non-zero vector $\mathbf{x}$, are an eigenvalue and a corresponding (right) eigenvector of M if they are such that $M\mathbf{x} = \lambda\mathbf{x}$. In the same way, if $\mathbf{x}^T M = \lambda\mathbf{x}$, $\mathbf{x}$ is called left eigenvector corresponding to the eigenvalue λ. Note that, a left eigenvector is a (right) eigenvector of the transpose matrix. A matrix is row-stochastic if its rows are non negative and the sum of each row is one. In this case, it is easy to show that there exists a dominant eigenvalue equal to 1 and a corresponding eigenvector $\mathbf{x} = (c, c, \ldots, c)^T$, for any constant c. A very simple method for the computation of the dominant eigenpair is the Power method [12] which, for stochastic irreducible matrices, is convergent for any choice of the starting vector with non negative entries. A stochastic matrix M can be viewed as a transition matrix associated to a family of Markov chains, where each entry M_{ij} represents the probability of a transition from state i to state j. By the Ergodic Theorem for Markov chains [24] an irreducible stochastic matrix M has a unique steady state distribution, that is a vector π such that $\pi^T M = \pi^T$. This means that the stationary distribution of a Markov chain can be determined by computing the left eigenvector of the stochastic matrix M. Given a graph $G = (V, E)$ and its adjacency matrix A, let outdeg(i) be the out-degree of vertex i that is the number of non-zeros in the i-th row of A. A node with no out-links is called "dangling".

2 Google's PageRank Model

In this section we review the original idea of Google's PageRank [4]. The Web is viewed as a directed graph (the Web Graph) $G = (V, E)$, where each of the N pages is a node and each hyperlink is an arc. The intuition behind this model is that a page $i \in V$ is "important" if it is pointed by other pages which are in turn "important". This definition suggests an iterative fixed-point computation to assigning a rank of importance to each page in the Web. Formally, in the original model [23], a random surfer sitting on the page i can jump with equal probability $p_{ij} = 1/\text{outdeg}(i)$ to each page j adjacent to i. The iterative equation for the computation of the PageRank vector $\mathbf{z}$ becomes $z_i = \sum_{j \in I_i} p_{ji} z_j$, where I_i is the set of nodes in-linking to the node i. The component z_i is the "ideal" PageRank of page i and it is then given by the sum of PageRank's assigned to the nodes pointing to i, weighted by the transition probability p_{ij}. The equilibrium distribution of each state represents the ratio between the number of times the random walks is in the state over the total number of transitions, assuming the random walks continues for infinite time. In matrix notation, the above equation is equivalent to the solution of the following system of equations $\mathbf{z}^T = \mathbf{z}^T P$, where $P_{ij} = p_{ij}$. This means that the PageRank vector $\mathbf{z}$ is the left eigenvector of P corresponding to the eigenvalue 1. In the rest of the paper, we assume that $||\mathbf{z}||_1 = \sum_{i=1}^{N} z_i = 1$, since the computation is not interested in assigning an exact value to each z_i, but rather in the relative rank between the nodes.

The "ideal" model has unfortunately two problems. The first problem is due to the presence of dangling nodes. They capture the surfer indefinitely. Formally, a dangling node corresponds to an all-zero row in P. As a consequence, P is not stochastic and the Ergodic Theorem cannot be applied. A convenient solution to the problem of dangling nodes is to define a matrix $\bar{P} = P + D$, where D is the rank one matrix defined as $D = \mathbf{dv}^T$, and $d_i = 1$ iff $\text{outdeg}(i) = 0$. The vector $\mathbf{v}$ is a *personalization vector* which records a generic surfer's preference for each page in V [14, 16]. The matrix $\bar{P}$ imposes a random jump to every other page in V whenever a dangling node is reached. Note that the new matrix $\bar{P}$ is stochastic. In Section 3, we refer this model as the "natural" model and compare it with other approaches proposed in the literature. The second problem, with the "ideal" model is that the surfer can "get trapped" by a cyclic path in the Web Graph. Brin and Page [4] suggested to enforce irreducibility by adding a new set of artificial transitions that with low probability jump to all nodes. Mathematically, this corresponds to defining a matrix $\widehat{P}$ as

$$\widehat{P} = \alpha \bar{P} + (1 - \alpha) \, \mathbf{ev}^T, \tag{1}$$

where $\mathbf{e}$ is the vector with all entries equal to 1, and α is a constant, $0 < \alpha < 1$. At each step, with probability α a random surfer follows the transitions described by $\bar{P}$, while with probability $(1 - \alpha)$ she/he bothers to follows links and jumps to any other node in V accordingly to the personalization vector $\mathbf{v}$. The matrix $\widehat{P}$ is stochastic and irreducible and both these conditions imply that the PageRank vector $\mathbf{z}$ is the unique steady state distribution of the matrix $\widehat{P}$ such that

$$\mathbf{z}^T \widehat{P} = \mathbf{z}^T. \tag{2}$$

From (1) it turns out that the matrix $\widehat{P}$ is explicitly

$$\widehat{P} = \alpha(P + \mathbf{dv}^T) + (1 - \alpha) \, \mathbf{ev}^T. \tag{3}$$

The most common numerical method to solve the eigenproblem (2) is the Power method [12]. Since $\widehat{P}$ is a rank one modification of αP, it is possible to implement a power method which multiplies only the sparse matrix P by a vector and upgrades the intermediate result with a constant vector, at each step.

The eigenproblem (2) can be rewritten as a linear system. By substituting (3) into (2) we get $\mathbf{z}^T(\alpha P + \alpha \mathbf{dv}^T) + (1 - \alpha)\mathbf{z}^T \mathbf{ev}^T = \mathbf{z}^T$, which means that the problem is equivalent to the solution of the following linear system of equations

$$S\mathbf{z} = (1 - \alpha)\mathbf{v}, \qquad (4)$$

where $S = I - \alpha P^T - \alpha \mathbf{vd}^T$, and we make use of the fact that $\mathbf{z}^T \mathbf{e} = \sum_{i=1}^{N} z_i = 1$. The transformation of the eigenproblem (2) into (4) opens the route to a large variety of numerical methods not completely investigated in literature. In next section we present a lightweight solution to handle the non-sparsity of S.

3 A Sparse Linear System Formulation

In this section we show how we can compute the PageRank vector as the solution of a sparse linear system. We remark that the way one handles the dangling node is crucial, since they can be a huge number[1]. Page et al. [23], adopted the drastic solution of removing completely the dangling nodes. Doing that, the size of the problem is sensibly reduced but a large amount of information present in the Web is ignored. This has an impact on both the dangling nodes - which are simply not ranked - and on the remaining nodes - which don't take into account the contribution induced by the random jump from the set of dangling nodes. Moreover, removing this set of nodes could potentially create new dangling nodes, which must be removed in turn.

Arasu et al. [1] handled dangling nodes in a different way respect to the natural model presented in Section 2. They modify the Web Graph by imposing that every dangling node has a self loop. In terms of matrices, $\bar{P} = P + F$ where $F_{ij} = 1$ iff $i = j$ and outdeg$(i) = 0$. The matrix $\bar{P}$ is row stochastic and the computation of PageRank is solved using a random jump similar to the equation (3), where the matrix F replaces D. This model is different from the natural model, as it is evident from the following example.

Example 1. Consider the graph in Figure 1 and the associated transition matrix. The PageRank obtained, by using the natural model, orders the node as $(2, 3, 5, 4, 1)$. Arasu's model orders the node as $(5, 4, 2, 3, 1)$. Note that in the latter case node 5 ranks better than node 2, which is not what one expects.

From the above observations we believe that it is important to take into account the dangling nodes and that the natural model is the one which better capture the behavior of a random surfer. The dense structure of the matrix S

---

[1] According to [19], a 2001 sample of the Web containing 290 million pages had only 70 million non-dangling nodes.

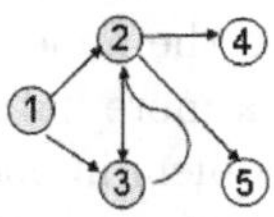

$$P = \begin{pmatrix} 0 & 1/2 & 1/2 & 0 & 0 \\ 0 & 0 & 1/3 & 1/3 & 1/3 \\ 0 & 1 & 0 & 0 & 0 \\ 0 & 0 & 0 & 0 & 0 \\ 0 & 0 & 0 & 0 & 0 \end{pmatrix}.$$

Fig. 1. An example of graph whose rank assignment differs if the dangling nodes are treated as in the model presented in [1]

poses serious problems to the solution of the linear system (4). On the contrary, the sparsity of the matrix P is easily exploited computing the PageRank vector with the Power method. In fact, it is common to implement the Power method in a way where the matrix-vector multiplications involve only the sparse matrix P while the rank-one modifications are handled separately [13].

In the following, we show how to manage dangling nodes in a direct and lightweight manner which makes it possible to use iterative methods for linear systems. In particular, we prove formally the equivalence of (4) to the solution of a system involving only the sparse matrix $R = I - \alpha P^T$.

Theorem 1. *The PageRank vector $\mathbf{z}$ solution of (4) is obtained by solving the system $R\mathbf{y} = \mathbf{v}$ and taking $\mathbf{z} = \mathbf{y}/\|\mathbf{y}\|_1$.*

Proof. Since $S = R - \alpha\,\mathbf{vd}^T$ equation (4) becomes $(R - \alpha\,\mathbf{vd}^T)\mathbf{z} = (1 - \alpha)\,\mathbf{v}$, that is, a system of equations where the coefficient matrix is the sum of a matrix R and a rank-one matrix. Note that R is non singular since $\alpha < 1$ and therefore all the eigenvalues of R are different from zero. We can use the Sherman-Morrison formula [12], paragraph 2.1.3, for computing the inverse of the rank-one modification of R. As a consequence, we have

$$(R - \alpha\mathbf{vd}^T)^{-1} = R^{-1} + \frac{R^{-1}\mathbf{vd}^T R^{-1}}{1/\alpha + \mathbf{d}^T R^{-1}\mathbf{v}}. \tag{5}$$

From (5), denoting by $\mathbf{y}$ the solution of the system $R\mathbf{y} = \mathbf{v}$, we have

$$\mathbf{z} = (1 - \alpha)\left(1 + \frac{\mathbf{d}^T\mathbf{y}}{1/\alpha + \mathbf{d}^T\mathbf{y}}\right)\mathbf{y},$$

that means that $\mathbf{z} = \gamma\mathbf{y}$, and the constant $\gamma = (1 - \alpha)\left(1 + \frac{\mathbf{d}^T\mathbf{y}}{1/\alpha + \mathbf{d}^T\mathbf{y}}\right)$ can be computed normalizing $\mathbf{y}$ in such a way $\|\mathbf{z}\|_1 = 1$.   □

Summarizing, we have shown that in order to compute the PageRank vector $\mathbf{z}$ we can solve the system $R\mathbf{y} = \mathbf{v}$, and then normalize $\mathbf{y}$ to obtain the PageRank vector $\mathbf{z}$. This means that the rank one matrix D in the PageRank model accounting for dangling pages plays a role only in the scaling factor γ. Moreover, the computation of γ is not necessary as well, since generally we are only interested in the relative ranks of the Web pages. Note that the matrix used by Arasu and et al. [1] is also sparse due to the way they deal with the dangling nodes, but the PageRank obtained don't ranks the node in a natural way (see Example 1). Instead, our approach guarantees a more natural ranking and handles the density of S by transforming a dense problem in one which uses the sparse matrix R. Moreover, when solving a linear system, particular attention must be devoted to the conditioning of the problem. It is easy to show [20, 17], that the condition number in the 1-norm of S is $\text{cond}_1(S) = \frac{1+\alpha}{1-\alpha}$, which means that the problem tends to become ill-conditioned as α goes to one. On the other hand, as we will show in the full paper, $\text{cond}_1(R) \le \text{cond}_1(S)$. However, on small synthetic matrices as well as on the web crawler of 24 million nodes used in the experimentation, we observed that the system involving R is much better conditioned than the one involving S for α going to 1. Bianchini and al. in [2] prove that the iterative method derived by (2) and involving $\widehat{P}$ produces the same sequence of vectors of the Jacobi method applied to matrix R. Recently [11], another way to deal with dangling nodes has been proposed. That approach is however different from ours. In particular, they assign separatelly a rank to dangling and non dangling pages and their algorithm requires the knowlege of a complete strongly connected subgraph of the web.

4   Exploiting the Web Matrix Permutations

In the previous section we have shown how to transform the linear system involving the matrix S into an equivalent linear system, where the coefficient matrix R is as sparse as the Web Graph. To solve the linear system $R\mathbf{y} = \mathbf{v}$ convenient strategies are Jacobi, Gauss-Seidel [25], because they use space comparable to that used by the Power method. These methods are convergent if and only if the spectral radius of the iteration matrix is strictly lower than one. Since R is an M-matrix, it is possible to show that both Jacobi and Gauss-Seidel methods are convergent and moreover that Gauss-Seidel method applied to R is always faster than Jacobi method [25]. When solving a sparse linear system a common practice [10] is to look for a reordering scheme that reduces the (semi)bandwidth for increasing data locality and hence the time spent for each iteration. For some methods, we also have a change in the spectral radius of the iteration matrix when applied to the permuted matrix. This is not the case for Jacobi method since the spectral radius of the iteration matrix is invariant under permutation. In the same way, the convergence of the Power method is also independent of matrix reordering. However, in [19] is given a method to reorder the matrix sorting the Urls lexicographically. This may help to construct a better starting vector for the Power method and to improve data locality. A much challenging perspective is reordering the Web matrix for the Gauss-Seidel method, where opportune permutations can lead both to an increase in data locality and to an iteration matrix with a reduced spectral radius.

Our permutation strategies are obtained by combining different elementary operations. A very effective reordering scheme, denoted by $\mathcal{B}$, is the one given by permuting the nodes of the Web graph according to the order induced by a

BFS visit. The BFS visit makes it possible to discover reducibility of the Web Matrix, since this visit assigns contiguous permutation indices to pages pointed by the same source. Therefore, this permutation produces lower block triangular matrices. It has been observed [9] that the BFS strategy for reordering sparse symmetric matrices produces a reduced bandwidth when the children of each node are inserted in order of decreasing degree. For this reason, we examine other reordering schemes which are obtained by sorting the nodes in terms of their degree. Since the Web matrix is not symmetric, we consider the permutation which reorders the pages for decreasing out-degree; denoting this scheme as $\mathcal{O}$ while the permutation $\mathcal{Q}$ sorts the pages of the Web matrix for increasing out-degree. Note that these permutations list the dangling pages in the last and in the first rows of the Web matrix respectively. We experimented also the permutations obtained reordering the matrix by increasing and decreasing in-degree; denoted by $\mathcal{X}$ and $\mathcal{Y}$ respectively.

Since $R = I - \alpha P^T$, our algorithms needs to compute the transpose of the Web matrix. We denote the transposition of the Web matrix by $\mathcal{T}$. The various operations can be combined obtaining 15 different possible reordering of the Web matrix as shown in Figure 2. In accordance with the taxonomy in Figure 2, we

Full	Lower Block Triangular	Upper Block Triangular
$\mathcal{T}$	$\mathcal{TB}$	$\mathcal{BT}$
$\mathcal{OT}$	$\mathcal{OTB}$	$\mathcal{OBT}$
$\mathcal{QT}$	$\mathcal{QTB}$	$\mathcal{QBT}$
$\mathcal{XT}$	$\mathcal{XTB}$	$\mathcal{XBT}$
$\mathcal{YT}$	$\mathcal{YTB}$	$\mathcal{YBT}$

Fig. 2. Web Matrix Permutation Taxonomy

denote, for instance, by $R_{\mathcal{XTB}} = I - \alpha \Pi(P)$, where the permuted matrix $\Pi(P)$ is obtained applying first the $\mathcal{X}$ permutation, then transposing the matrix and applying finally the $\mathcal{B}$ permutation on the matrix reordered. The first column in figure 2 gives rise to full matrices, while the second and third columns produce block triangular matrices due to the BFS's order of visit. In Figure 3, we show a plot of the structure of a Web matrix rearranged according to each item of the above taxonomy. We adopted ad hoc numerical methods for dealing with the different shapes of matrices in Figure 3. In particular, we compared Power method, and Jacobi iterations with Gauss-Seidel, and Reverse Gauss-Seidel. We recall that the Gauss-Seidel method computes $y_i^{(k+1)}$, the i−th entry of the vector at the $(k + 1)$−th iteration step as a linear combination of $y_j^{(k+1)}$ for $j = 1, \ldots, i - 1$ and of $y_j^{(k)}$ for $j = i + 1, \ldots, n$. On the contrary, the Reverse Gauss-Seidel method computes the entries of the vector $\mathbf{y}^{(k+1)}$ bottom up, that is it computes $y_i^{(k+1)}$ for $i = n, \ldots, 1$ as a linear combination of $y_j^{(k+1)}$ for $j = n, \ldots, i + 1$ and of $y_j^{(k)}$ for $j = 1, \ldots, i - 1$. Note that $R_{\mathcal{OT}} = J R_{\mathcal{QT}} J^T$ and $R_{\mathcal{XT}} = J R_{\mathcal{YT}} J^T$ where J is the anti-diagonal matrix, that is $J_{ij} = 1$ iff

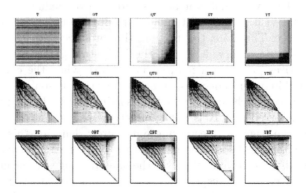

Fig. 3. Different shapes obtained rearranging the Web matrix P in accordance to the taxonomy. First row represents full matrices; second and third lower and upper block triangular matrices. Web Graph is made of 24 million nodes and 100 million links

$i+j = n+1$. This means that applying Gauss-Seidel to $R_{\mathcal{O}\mathcal{T}}$ ($R_{\mathcal{X}\mathcal{T}}$) is the same that applying Reverse Gauss-Seidel to $R_{\mathcal{Q}\mathcal{T}}$ ($R_{\mathcal{Y}\mathcal{T}}$).

The shapes of some matrices in Figure 3, encourage to exploit the matrix reducibility experimenting with block methods. In particular, we note that the matrices permuted according to the BFS visit have a block triangular structure. Moreover, also the matrix $R_{\mathcal{O}\mathcal{T}}$ is lower block triangular, since it separates non-dangling nodes from dangling nodes. A natural way to handle block triangular matrices is to use forward/backward block substitution. For instance, on the lower block triangular system

$$
\begin{bmatrix}
R_{11} & & & \\
R_{21} & R_{11} & & \\
\vdots & & \ddots & \\
R_{m1} & \cdots & & R_{mm}
\end{bmatrix}
\begin{bmatrix}
\mathbf{y}_1 \\
\mathbf{y}_2 \\
\vdots \\
\mathbf{y}_m
\end{bmatrix}
=
\begin{bmatrix}
\mathbf{v}_1 \\
\mathbf{v}_2 \\
\vdots \\
\mathbf{v}_m
\end{bmatrix},
$$

the solution can be computed as follows

$$
\begin{cases}
\mathbf{y}_1 = R_{11}^{-1}\mathbf{v}_1, \\
\mathbf{y}_i = R_{ii}^{-1}\left(\mathbf{v}_i - \sum_{j=1}^{i-1} R_{ij}\mathbf{y}_j\right) & \text{for } i = 2,\ldots,m.
\end{cases}
$$

This requires the solution of m smaller linear systems, where the coefficient matrices are the diagonal blocks in the order they appear. As solving methods for the diagonal block systems, we tested both Gauss-Seidel and Reverse Gauss-Seidel methods. We denote by LB and UB the methods obtained using Gauss-Seidel as solver of the diagonal blocks on lower or upper block structures respectively. LBR and UBR use instead Reverse Gauss-Seidel to solve the diagonal linear systems. Summing up, we have the taxonomy of solution strategies

Scalar methods	shapes	Block methods	shapes
PM	all	LB	R_{*TB} and R_{OT}
Jac	all	LBR	R_{*TB} and R_{OT}
GS	all	UB	R_{*BT}
RGS	all	UBR	R_{*BT}

Fig. 4. Numerical Methods Taxonomy. PM is the Power method, Jac denotes the Jacobi method, GS and RGS are the Gauss-Seidel and Reverse Gauss-Seidel, respectively. All of them can be applied to each transformation of the matrix according to the taxonomy 2. Among block-methods we have LB and LBR which can be applied to all lower block triangular matrices and use GS or RGS to solve each diagonal block. Similarly, UB and UBR refer to the upper block triangular matrices

reported in figure 4. In Section 5 we report the experimental results obtained by applying each method in Figure 4 to all the suitable matrices in Figure 2.

5   Experimental Results

We tested the approaches discussed in previous sections using a Web Graphs obtained as a crawling of 24 million Web pages with about 100 million hyper-links and containing approximately 3 million dangling nodes. This data set was donated to us by the Nutch project (see http://www.nutch.org/). We run our experiments on a PC with a Pentium IV 3GHz, 2.0GB of memory and 512Mb of L1 cache. A stopping criterion of 10^{-7} is imposed on the absolute difference between the vectors computed in two successive iterations. In Figure 5 we report the running time in seconds, the Mflops and the number of iterations for each combination of solving and reordering methods described in 4 and 2. Some cells are empty since there are methods suitable only on particular shapes. Moreover, in Figure 5 the results in the last two columns are relative to LB and LBR meth-ods for lower block triangular matrices and UB or UBR for upper block triangular matrices. We do not report in the table the behavior of Jac method, since it has always worst performance than GS method. Since the diagonal entries of R are equal to 1, Jacobi method is essentially equivalent to the Power method. In our case, the only difference is that Jac is applied to R while PM works on $\widehat{P}$, which incorporates the rank one modification accounting for dangling nodes. We implemented PM using the optimizations suggested in [23, 13]. Using the results of Theorem 1, we get a reduction in Mflops of about 3%. For the increased data locality, the running time of Jac benefits from matrix reordering, and we have a reduction up to 23% over the Power iterations.

We now compare the proposed methods versus PM applied to the original matrix since this is the common solving method to compute PageRank vector. Other comparisons can be obtained from Figure 5. As one can expect, the use of GS and RGS on the original matrix already accounts for a reduction of about 40% in the number of Mflops and of about 45% in running time. These improvements

Name	PM	GS	RGS	LB/UB	LBR/UBR
T	3454 33093 152	1903 19957 92	2141 20391 94	$- - -$	$- - -$
OT	2934 33093 152	1660 20825 85	1784 19957 104	1570 19680 96	1515 18860 92
XT	3315 33309 153	1920 21259 98	1944 19957 92	$- - -$	$- - -$
TB	1386 32876 151	731 21910 101	717 18439 85	485 16953 100	438 15053 82
OTB	1383 33093 152	705 21476 99	705 18439 85	480 17486 97	446 15968 85
QTB	1353 32876 151	743 23645 109	618 16920 78	520 18856 106	**385 13789 75**
XTB	1361 33309 153	682 21259 98	715 19090 88	484 17196 98	472 16414 87
YTB	1392 32876 151	751 22343 105	**625 16270 75**	501 17972 100	390 13905 75
BT	1394 33093 152	628 18439 85	879 22560 104	464 15545 85	570 19003 104
OBT	1341 33309 153	605 18873 87	806 21693 100	470 15937 87	543 18312 100
QBT	1511 33093 152	702 18873 87	922 21693 100	427 15128 87	493 17387 100
XBT	1408 33093 152	667 19306 89	860 21693 100	474 16075 88	541 18265 100
YBT	1351 33093 152	600 18439 85	806 21693 100	461 15564 85	541 18310 100

Fig. 5. Experimental Results: in columns are listen the numerical methods analyzed and the rows describe the permutations applied to the matrix R. Each cell represents the running time in seconds, the number of Mflops and the number of iterations taken by the solving methods. Note that the results in the last two columns account either for the cost of the LB and LBR methods, applied to lower block triangular matrices, or for the cost of UB and UBR methods, applied to upper block triangular matrices. In bold we highlight our best results for scalar and block methods

are striking when the system matrix is permuted. The best performance of scalar methods is obtained using the YTB combination of permutations on RGS method. This yields a Mflops reduction of 51% with respect to PM and a further reduction of 18% with respect to the GS both applied to the full matrix. The running time is reduced of 82%.

The common intuition is that Gauss-Seidel method behaves better on a quasi-lower triangular while Reverse Gauss-Seidel is faster when applied to quasi-upper triangular matrices. However, in this case the intuition turns to be misleading. In fact, for our Web matrix RGS works better on lower block triangular matrices and GS works better on upper matrices. Even better results are obtained by block methods. LB applied to R_{OT} achieves a reduction of 41% in Mflops with respect to PM. Adopting this solving method, we explore just the matrix reducibility due to dangling nodes. As depicted in Figure 6, the best result is obtained for the QTB permutation when the LBR solving method is applied. In this case, we have a reduction of 58% in Mflops and of 89% in the running time. This means that our solving algorithm computes the PageRank vector in about a tenth of the running time and with less than half operations of the Power method. The results given in Figure 5 do not take into account the effort spent in reordering the matrix. However, the most costly reordering scheme is the BFS visit of the Web graph, which can be efficiently implemented in semi-external memory as reported in [6, 22]. The running time spent for doing the BFS are comparable to those reported in [5], where less of 4 minutes are taken on a Web Graph with 100

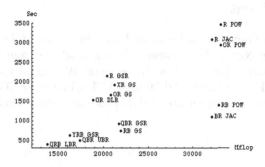

Fig. 6. A plot of some of the results of Figure 5. On the x-axis the number of Mflops and on the y-axis the running time in seconds. Each point is labeled with the permutation applied and the solving method used

million nodes and it is however largely repaid from the speedup achieved on the solving methods. Moreover, in case of personalized PageRank the permutations can be applied only once and reused for all personalized vectors **v**. An intuitive picture of the gain obtained by combining permutation strategies with the scalar and block solving methods is shown in Figure 6.

6   Conclusion

The ever-growing size of Web graph implies that the value and importance of fast methods for Web ranking is going to rise in the future. The problem of PageRank computation can be easily be viewed as a dense linear system. We showed how to handle the density of this matrix by transforming the original problem in one which uses a matrix as sparse as the Web Graph itself. On the contrary of what done in [1], we achieved this result without altering the original model. This result allows to efficiently consider the PageRank computation as a sparse linear system, in alternative to the eigenpairs interpretation. Dealing with a sparse linear system opens the way to exploit the Web Matrix reducibility by composing opportunely some Web matrix permutations to speedup the PageRank computation. We showed that permuting the Web matrix according to a combination of in-degree or out-degree and sorting the pages following the order of the BFS visit, can effectively increase data locality and reduce the running time when used in conjunction with numerical method such as lower block solvers. Our best result achieves a reduction of 58% in Mflops and of 89% in terms of seconds required compared to the Power method commonly used to compute the PageRank. This means that our solving algorithm requires almost a tenth of the time and much less than half in terms of Mflops. The previous better improvement over the Power method is due to [21] where a reduction of 80% in time is achieved on a data set of roughly 400.000 nodes. In light of the experimental results, our approach for speeding up PageRank computation appears much promising.

Acknowledgment

We thank Daniel Olmedilla of Learning Lab Lower Saxony, Doug Cutting and
Ben Lutch of the Nutch Organization, Sriram Raghavan and Gary Wesley of
Stanford University. They provided to us some Web graphs and a nice Web
crawler. We thank Luigi Laura and Stefano Millozzi for their COSIN library. We
also thank Paolo Ferragina for useful discussions and suggestions.

References

1. A. Arasu, J. Novak, A. Tomkins, and J. Tomlin. PageRank computation and the
   structure of the Web: Experiments and algorithms. In *Proc. of the 11th WWW
   Conf.*, 2002.
2. M. Bianchini, M. Gori, and F. Scarselli. Inside PageRank. *ACM Trans. on Internet
   Technology*, 2004. to appear.
3. P. Boldi and S. Vigna. WebGraph framework i: Compression techniques. In *Proc.
   of the 23th Int. WWW Conf.*, 2004.
4. S. Brin and L. Page. The anatomy of a large-scale hypertextual Web search engine.
   Computer Networks and ISDN Systems, 30(1–7):107–117, 1998.
5. A. Z. Broder, R. Kumar, F. Maghoul, P. Raghavan, S. Rajagopalan, R. Stata,
   A. Tomkins, and J. L. Wiener:. Graph structure in the Web. *Computer Networks*,
   33:309–320, 2000.
6. A. L. Buchsbaum, M. Goldwasser, S. Venkatasubramanian, and J. Westbrook. On
   external memory graph traversal. In *SODA*, pages 859–860, 2000.
7. Y. Chen, Q. Gan, and T. Suel. I/o-efficient techniques for computing Pagerank.
   In *Proc. of the 11th WWW Conf.*, 2002.
8. J. Cho and S. Roy. Impact of Web search engines on page popularity. In *Proc. of
   the 13th WWW Conf.*, 2004.
9. E. Cuthill and J. McKee. Reducing the bandwidth of sparse symmetric matrices.
   In *Proc. 24th Nat. Conf. ACM*, pages 157–172, 1969.
10. C. Douglas, J. Hu, M. Iskandarani, M. Kowarschik, U. Rüde, and C. Weiss. Max-
    imizing cache memory usage for multigrid algorithms. In *Multiphase Flows and
    Transport in Porous Media: State of the Art*, pages 124–137. Springer, 2000.
11. N. Eiron, S. McCurley, and J. A. Tomlin. Ranking the web frontier. In *Proc. of
    13th WWW Conf.*, 2004.
12. G. H. Golub and C. F. Van Loan. *Matrix Computations.* The John Hopkins
    University Press, Baltimore, 1996. Third Edition.
13. T. Haveliwala. Efficient computation of PageRank. Technical report, Stanford
    University, 1999.
14. T. Haveliwala. Topic-sensitive PageRank. In *Proc. of the 11th WWW Conf.*, 2002.
15. T. Haveliwala, S. Kamvar, and G. Jeh. An analytical comparison of approaches to
    personalizing PageRank. Technical report, Stanford University, 2003.
16. G. Jeh and J. Widom. Scaling personalized Web search. In *Proc. of the 12th
    WWW Conf.*, 2002.
17. S. Kamvar and T. Haveliwala. The condition number of the pagerank problem.
    Technical report, Stanford University, 2003.
18. S. Kamvar, T. Haveliwala, C. Manning, and G. Golub. Extrapolation methods for
    accelerating PageRank computations. In *Proc. of 12th. WWW Conf.*, 2003.

19. S. D. Kamvar, T. H. Haveliwala, C. Manning, and G. H. Golub. Exploiting the block structure of the Web for computing PageRank. Technical report, Stanford University, 2003.
20. A. N. Langville and C. D. Meyer. Deeper inside PageRank. *Internet Mathematics*, 2004. to appear.
21. C. P. Lee, G. H. Golub, and S. A. Zenios. A fast two-stage algorithm for computing PageRank. Technical report, Stanford University, 2003.
22. K. Mehlhorn and U. Meyer. External-memory breadthfirst search with sublinear I/O. In *European Symposium on Algorithms*, pages 723–735, 2002.
23. L. Page, S. Brin, R. Motwani, and T. Winograd. The PageRank citation ranking: Bringing order to the Web. Technical report, Stanford, 1998.
24. W. S. Stewart. *Introduction to the Numerical Solution of Markov Chains*. Princeton University Press, 1995.
25. R. S. Varga. *Matrix Iterative Analysis*. Prentice-Hall, Englewood Cliffs, 1962.

T-Rank: Time-Aware Authority Ranking

Klaus Berberich[1], Michalis Vazirgiannis[1,2], and Gerhard Weikum[1]

[1] Max-Planck Institute of Computer Science,
Saarbruecken, Germany
{kberberi, mvazirgi, weikum}@mpi-sb.mpg.de
[2] Dept of Informatics, Athens Univ. of Economics and Business,
Athens, Greece
mvazirg@aueb.gr

Abstract. Analyzing the link structure of the web for deriving a page's authority and implied importance has deeply affected the way information providers create and link content, the ranking in web search engines, and the users' access behavior. Due to the enormous dynamics of the web, with millions of pages created, updated, deleted, and linked to every day, timeliness of web pages and links is a crucial factor for their evaluation. Users are interested in important pages (i.e., pages with high authority score) but are equally interested in the recency of information. Time – and thus the freshness of web content and link structure - emanates as a factor that should be taken into account in link analysis when computing the importance of a page. So far only minor effort has been spent on the integration of temporal aspects into link analysis techniques. In this paper we introduce T-Rank, a link analysis approach that takes into account the temporal aspects freshness (i.e., timestamps of most recent updates) and activity (i.e., update rates) of pages and links. Preliminary experimental results show that T-Rank can improve the quality of ranking web pages.

1 Introduction

The web graph grows at a tremendous rate while its content is updated at a very high pace following interesting patterns [8]. The issue of analyzing the link structure of the web, in order to derive a page's authority has attracted significant efforts [12, 17] and has affected the way web information providers create and link content and the users' behavior in searching and browsing information.

The predominant authority ranking approach in search engines is the PageRank method [17]. However, this is a pre-computed static measure that does not reflect the timeliness or temporal evolution of pages and links. Due to the high dynamics of the web, with millions of pages created, updated, deleted, and linked to every day, timeliness of web pages and links is a crucial factor for their evaluation. Users are interested not just in the most important pages (i.e., the ones with the highest authority), but in information that is both important and timely. For example, the query "olympics opening" based on PageRank returns among its highest ranked authorities pages about the 2002 winter Olympics in Salt Lake City, but if the query is

S. Leonardi (Ed.): WAW 2004, LNCS 3243, pp. 131–142, 2004.

asked in spring 2004 the user is more likely to be interested in the upcoming summer Olympics in Athens whose web pages, however, have not yet gained such a high static PageRank to qualify for the very top results.

Often the temporal focus of a user would be on the present time, then the overall ranking should reflect the time of the last updates to a page and its in-links, called the freshness in this paper. A second aspect that should affect the importance of a page is the rate of updates, called activity in this paper, both for its content and for its in-links. It is intuitive that pages whose content or in-links are frequently updated will be of greater interest to users. In contrast, pages with no updates and "frozen" in-links indicate lack of activity and interest from the web community.

In addition to focusing on specific time points, a user may be interested in the important pages within a temporal window of interest (twi) referring to a time period such as the years 1998 through 1999 with the anxiety about the Y2K problem. In such a case it seems appropriate to interpret the twi in a relaxed sense, taking into account also information that slightly falls outside the twi (e.g., the first months of 2000 and the year 1997) but with reduced weight only.

The seminal work of Brin and Page [17] and Kleinberg [12] entailed rich work on link analysis and authority ranking for the Web [3, 4, 5, 9, 10, 16, 19]. However, only minor effort has been spent on the integration of temporal aspects into ranking techniques for web or intranet search. Amitay et al. [3] state that if a page's last modification date is available then search engines will be able to provide results that are more timely and better reflect current real-life trends. Kraft et al. [13] provide a formal description of the evolving web graph. Also they describe their approach towards extracting the evolving web graph from the Internet Archive [2] and present statistics on the extracted data. Another related effort by Baeza-Yates et al. [4] considers the age and rate of change of web pages and their effect on the Web structure. Finally, the web graph and its evolution have been the subject of several recent studies [6, 8, 13, 14, 15]. None of the prior work addresses the technical issue of how to factor time into the importance ranking of web pages based on link analysis.

The T-Rank method developed in this paper extends link analysis by taking into account temporal aspects like freshness and activity of both pages and links. Technically, our approach can be seen as an extension of the PageRank-style family of random-walk models. Thus, just like PageRank we compute stationary probabilities of pages for an appropriately defined Markov chain (by numerical power iteration). Our key innovation is to introduce time-awareness and page/link activity into the random walk and the resulting Markov chain. We present a framework that can be specialized in different ways by setting model parameters in the desired way.

Preliminary experiments indicated that this approach can indeed significantly improve the ranking of web search results with temporal focus. Our techniques are not limited to the web but can be applied to any kind of graph structure. In one experiment we use the DBLP authors' bibliometric network, with authors as nodes and citations as edges. A second experiment is based on crawls of product web pages at amazon.com.

2 Temporal Model for Link Analysis

The building blocks of the formal model are a directed graph with temporal annotations and a specification of a temporal focus of attention. The directed graph $G(V,E)$ consists of the set of nodes V and the set of edges E. Both nodes and edges in the graph are annotated with timestamps indicating different kinds of events. Assuming that the timestamps are represented by natural numbers, time is seen to be discretized at some granularity. For each node and each edge, one timestamp $TS_{creation}$ incorporates the moment of creation. Analogously the timestamp $TS_{deletion}$ represents the moment of deletion. Furthermore $TS_{modifications}$ refers to the set of timestamps representing the points when the object was modified. We define $TS_{lastmod}$ to represent the most recent timestamp in the set $TS_{modifications}$.

Another part of the model is the *temporal focus of attention* representing the period that the user is interested in. It is represented by a *temporal window of interest* (*twi*) defined by two timestamps TS_{origin} and TS_{end}, where $TS_{origin} < TS_{end}$. In the case $TS_{origin} = TS_{end}$ the user is interested in a single timestamp (further referred to as TS_{focus}). As mentioned above we provide a "tolerance" interval that extends *twi* with the parameters *t1* and *t2* (see Figure 1), that fulfills: $t1 \leq TS_{origin} \leq TS_{end} \leq t2$.

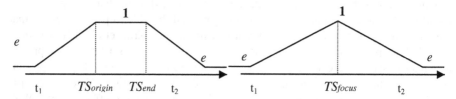

Fig. 1. Temporal proximity window for a temporal interval

Fig. 2. Temporal proximity window for a temporal point

Here we assume that a page's importance within the *twi* is maximum (set to the value 1) while it decreases linearly to lower values outside of *twi* with its *temporal proximity* to *twi*. We have chosen this linear approach for simplicity; other forms of weighting functions would be possible, too. In the case that the *twi* is reduced to a single timestamp of interest (TS_{focus}) the evolution of importance vs. time is depicted in Figure 2.

The auxiliary parameters t1, t2 are defined by the user-specific system setup and allow us to cover different application requirements. Assume, for instance, that a user is interested in the period *after an unexpected event* (e.g., the unfortunate terrorism event in Madrid on March 11, 2004). In this case we set TS_{focus} = "March 11, 2004" while there is no need to compute importance for pages before this timestamp as there were no relevant pages to this event. This implies $t1 = TS_{focus}$ while $t2 > TS_{focus}$ (see Fig. 3 (a)). We can also handle the case where a user's temporal focus of interest excludes the period before (or after) the relevant *twi* (e.g., a user interested in the three-weeks period of the "Tour de France 2003" and its discussion in the following three months, see Figure 3 (b)).

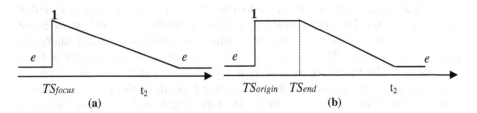

Fig. 3. Different cases of temporal interest

Assuming a *twi* $[TS_{origin}, TS_{end}]$ and a timestamp *ts* the *freshness* $f(ts)$ of *ts* with regard to *twi* is maximum (set to 1 if *ts* $\in$ *twi*) and decreasing towards small values as *ts* moves away from TS_{end} or TS_{origin}. To ensure convergence of T-Rank computations, we do not let freshness converge to zero, but rather assume that there is a minimum freshness value $e > 0$, and this is the value $f(ts)$ outside of the interval *[t1, t2]*. (You may think of this trick as a smoothing technique.) Then the complete definition of freshness is as follows:

$$f(ts) = \begin{cases} if\ Ts_{origin} \leq ts \leq Ts_{end}: & 1 \\ if\ t_1 \leq ts < Ts_{origin}: & \dfrac{1}{(TS_{origin} - ts) + 1} \\ if\ TS_{end} < ts \leq t_2: & \dfrac{1}{(ts - TS_{end}) + 1} \\ otherwise: & e \end{cases} \tag{1}$$

The second key notion to be defined here is *activity*. The activity of an object is its frequency of change (update rate) within the temporal focus of attention. The basis for computing the activity of a page or link is the set of timestamps *TSmodifications*. The sum of the freshness values of the timestamps in this set measures the activity of the object. The same applies for the *twi* case. In both cases, only modifications between *t1* and *t2* will be considered. Note that this gives higher weight to updates closer to *t2*; so if two pages have the same number of updates within a *twi* but one has more recent updates, then this page is considered more active. In the case that in *[t1 ,t2]* there was no update then we set the activity value to a small constant $e' > 0$ to ensure convergence of the T-Rank computation.

Then the complete definition of *activity* for an object (i.e., page or link) *o* is as follows:

$$a(o) = \begin{cases} if\ TS_{modifications} \neq \varnothing: & \displaystyle\sum_{t1}^{t2} \{f(ts) | ts \in TS_{modifications}\} \\ otherwise: & e' \end{cases} \tag{2}$$

3 The T-Rank Method

The T-Rank method builds on and extends the PageRank technique [17] which is based on a random-walk model. A random surfer traverses the web graph by either following outgoing links or by randomly jumping to another node. In both cases the next page visited is chosen according to a uniform probability distribution among the possible target pages. The event of making a random jump occurs with a fixed probability ε. The PageRank measure of a page is its stationary visit probability, or equivalently, the fraction of time the random surfer spends on this page as the length of the random walk approaches infinity. The following formula gives the PageRank r(y) of a page y:

$$r(y) = \sum_{(x,y)\in E} (1-\varepsilon)\frac{r(x)}{\text{outdegree}(x)} + \frac{\varepsilon}{n} \tag{3}$$

An interpretation of this recursive equation is that every page "distributes" its authority uniformly to all its successors, which is captured in the first term of the formula. This corresponds to the random surfer clicking on a link with probability $(1-\varepsilon)$ and choosing with uniform probability $1/\text{out deg ree}(x)$. The second term in the equation corresponds to the authority a node gains when being the target of a random jump.

The original PageRank work [17] already described the possibility of modifying the probability distribution underlying the random jump to favor certain nodes. The intention was to allow personalized or topic-sensitive authority scores. This idea has been adopted and developed further in [9, 10]. Xue et al. [19] describe another approach using non-uniform distributions for the transition probabilities. Our generalization of the PageRank method, which is described in the following formula, introduces a bias for both transition probabilities and random jump probabilities:

$$r(y) = \sum_{(x,y)\in E} (1-\varepsilon)\cdot t(x,y)\cdot r(x) + \varepsilon\cdot s(y) \tag{4}$$

Here the function $t(x, y)$ describes the transition probabilities. A specific value of $t(x, y)$ corresponds to the probability that a random surfer being in node x chooses to go to node y. The function $s(y)$ describes the random jump probabilities, a probability distribution which is independent of the random surfer's current position.

Casting Eq. 4 into matrix form and some simple transformations lead to:

$$\pi = M^T \cdot \pi \tag{5}$$

where the entries of column vector π correspond to the PageRank values pages. The matrix M is defined as $(1-\varepsilon)\cdot T + \varepsilon\cdot S$, where $T_{ij} = t(i, j)$ and $S_{ij} = s(j)$. Solving Eq. 5 for π is a Principal Eigenvector computation which can be numerically handled using the power iteration method. We choose an initial probability vector x_o (e.g., a uniform vector), and compute x_k, the result of the k-th iteration, by $x_k = M^T \cdot x_{k-1}$. The computation can be stopped, when the size of the residual vector $\|x_k - x_{k-1}\|_1$ is below a threshold δ.

Next, we define freshness and activity of pages and links affect the transition probabilities t(x, y) and the random jump probabilities s(y). Figure 4 illustrates the random surfer's situation with ts indicating the availability of timestamps. A random surfer with a particular temporal focus of attention who currently resides on some page x gives weights to the outgoing neighbors based on the temporal information developed in Section 2.

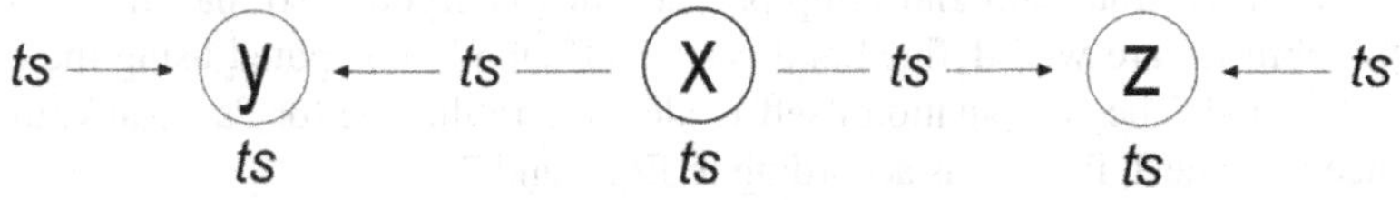

Fig. 4. Transition probabilities

On the web the temporal information for outgoing links is under the control of the source pages and therefore susceptible to "cheating". On the other hand the incoming links reflect the attention a web page has attracted and are more democratic in their nature and less susceptible to cheating. Among those incoming links the link emanating in the random surfer's current position can be picked out and treated in a special way. These observations suggest that the probability of the random surfer choosing y when leaving her current page x is a combination of the freshness f(y) of the target page y, the freshness f(x,y) of the link from x to y, and the average freshness of all incoming links of y. This leads to the following equation (where each of the three parts has a weight coefficient w_{ti} and the freshness values are normalized to lie between 0 and 1):

$$t(x, y) = w_{t1} \cdot \frac{f(y)}{\sum_{(x,z) \in E} f(z)} + w_{t2} \cdot \frac{f(x, y)}{\sum_{(x,z) \in E} f(x, z)} + w_{t3} \cdot \frac{avg\{f(v, y) \mid (v, y) \in E\}}{\sum_{(x,w) \in E} avg\{f(v, w) \mid (v, w) \in E\}} \tag{6}$$

The coefficients wt_i are considered as tunable parameters and must add up to 1. The function *avg* used here could easily be replaced by other aggregation functions.

The second building block of the random walk is the random jump. Analogously to our approach for defining temporally biased transition probabilities, we define desirable influence factors for choosing a page as a random jump target, namely, the freshness of a target page, the freshness of its incoming links, and the activity observed for nodes and edges. We recall that activity measures the frequency of modifications applied to nodes and edges. On the web, activity has a counterpart in the real world, as it seems to be common that users directly access highly updated web pages (e.g. a news site) and then subsequently follow a chain of links. These considerations suggest that the random jump probability of a target page y is a (weighted) combination of the freshness of y, the activity of y, the average freshness of the incoming links of y, and the average activity of the pages that link to y. This leads to the following equation (where the coefficients ws_i are again tunable parameters and have to add up to 1):

$$s(y) = w_{r1} \cdot \frac{f(y)}{\sum\limits_{z \in V} f(z)} + w_{r2} \cdot \frac{a(y)}{\sum\limits_{z \in V} a(z)} + w_{r3} \cdot \frac{avg\{f(v, y \mid (v, y) \in E\}}{\sum\limits_{z \in V} avg\{f(w, z \mid (w, z) \in E\}} + w_{r4} \cdot \frac{avg\{a(v, y \mid (v, y) \in E\}}{\sum\limits_{z \in V} avg\{a(w, z \mid (w, z) \in E\}} \quad (7)$$

It can be easily shown that the Markov chain defined by the random walk with temporally biased transition and jump probabilities is ergodic, so that the stationary state probabilities are well-defined and can be efficiently computed using the power iteration method. The computation itself is identical to the one for PageRank; the key difference is in the T-Rank bias according to Eq. 6 and 7.

4 Preliminary Experiments

We implemented the T-Rank method, including the preparation of time-annotated web or intranet pages, in Java (using Sun's JDSK 1.4.2) as a component of the BINGO! focused crawler [18] (a prototype system developed in our group).

All timestamps are encoded as natural numbers, allowing us to use compact integers in the database (as opposed to the data type Date which has considerably higher overhead).

We performed experiments on two different data sets: a bibliometric link graph derived from DBLP and a collection of product-related web pages of amazon.com. All our experiments were run on low-end commodity hardware (a 3-GHz Pentium PC with 1 GB of memory and a local IDE disk, under Windows XP).

4.1 The DBLP Experiment

The DBLP project [1] provides information on approximately half a million computer science publications with citation information available only for a subset. We constructed a graph structure from this data as follows: Authors are mapped to nodes in the graph; citations between (papers of) authors correspond to directed edges in the graph. The timestamp $TS_{creation}$ of a node corresponds to the date of the author's first publication. Similarly an edge between a source node A and a target node B is created when author A cited B for the first time. If A repeatedly cites B he/she causes the respective edge to be modified and thus creates a new element in $TS_{modifications}$. A node is modified for each new publication of an author. The resulting graph has 16.258 nodes with about 300.000 edges (as citations are captured only partly).

The timestamps of the resulting graph span an extended period (from about 1970 to now). There are similarities to the web graph with regards to the in-degree distribution that follows a power law of the form $P[\text{in-degree} = n] \sim (1/n)^k$ with $k > 0$.

We first performed a *baseline* experiment where we computed the T-Rank authority of nodes for different *twis*, each spanning one decade (1970-1980, 1980-1990, 1990-2000, 2000-now) with t1 and t2 set to two years preceding TS_{origin}

and following TS_{end}, respectively. The top-5 authors per decade are presented in Table 1, in comparison to static PageRank results for the entire period.[1]

Table 1. Tracing author importance in decades temporal intervals

	PageRank	T-Rank 70s	T-Rank 80s	T-Rank 90s	T-Rank 00s
1	E. F. Codd	E. F. Codd	E. F. Codd	E. F. Codd	Jim Gray
2	Donald D.Chamberlin	Raymond F. Boyce	Donald D.Chamberlin	Morton M. Astrahan	Jeffrey D. Ullman
3	Michael Stonebraker	H. Albrecht Schmid	Kapali P. Eswaran	Michael Stonebraker	Michael Stonebraker
4	Jim Gray	Edward M. McCreight	John Miles Smith	Irving L. Traiger	Jim Melton
5	Raymond A. Lorie	Donald D.Chamberlin	Morton M. Astrahan	Jim Gray	Hector Garcia-Molina

We then studied the *parameter tuning* of T-Rank by evaluating different assignments to the weights in the transition (wt_1, wt_2, wt_3) and the random jump (ws_1, ws_2, ws_3, ws_4) probabilities. Here the *twi* was invariantly set to $t1=1998$, $TS_{Origin}=1999$, $TS_{End}=2000$, $t2= 2001$. The remaining parameters were as follows: $\varepsilon =0.1$, $e=10^{-7}$, $\delta=10^{-9}$. Recall that the wt_i and ws_i weights reflect the user's bias regarding 1) the freshness of the target node, 2) the freshness of the individual links, 3) the aggregate freshness of all the incoming links of the target node. In our experiments the influence of the ws_i weights was marginal (probably because of the relatively small ε value); so we report only on results with all ws_i uniformly set to 0.25.

For each setting of the wt_i weights we compared the top-1000 results to the top-1000 results of the static PageRank, using the following metrics for comparing two top-k rankings over a set S of n objects (k < n):

1. the overlap similarity *Osim* [9], which is the number of objects that appear in both top-k sets,

2. the *Ksim* measure [9] based on Kendall's tau [7, 11], measuring the probability that two top-k rankings agree on the relative ordering of a randomly selected pair (u, v) where u,v belong to the union of the two top-k sets,

3. the *rank distance RD*, based on the footrule distance [7, 11], between two lists L1 and L2 as: $\dfrac{1}{n}\sum_{x \in S} |rank(x,L1) - rank(x,L2)|$ where in each list all objects that do not qualify for the top-k are mapped to a "virtual" rank k+1,

4. the *weighted rank distance WRD*, where the weight of an object's rank distance in two lists L1 and L2 is inversely proportional to the product of the two ranks: $$\frac{1}{n}\sum_{x \in S} \frac{| rank(x,L1) - rank(x,L2) |}{rank(x,L1)\cdot rank(x,L2)}.$$

The results are presented in Table 2. It is evident that T-Rank results differ significantly from those of static PageRank. For *WRD* an author's position in T-Rank

---

[1] These results must be interpreted with caution. The ranked lists do not necessarily capture the true scientific merits of authors, but merely reflect the extent to which authority is expressed in citations within the, fairly partial, citation information available at DBLP.

differs, on average (with the top ranks having appropriately higher weights) by 6 to 8 positions. Between the four different settings of the wt_i parameters, it is difficult to analyze the differences and arrive at conclusions for tuning. This aspect clearly requires further experimental studies.

Table 2. Top-1000 lists comparison for different T-Rank parameters

	PageRank	T-Rank1 $wt_1 = wt_2 = wt_3 = 1/3$	T-Rank2 $wt_1 = 1 wt_2 = wt_3 = 0$	T-Rank3 $wt_2 = 1 wt_1 = wt_3 = 0$	T-Rank4 $wt_3 = 1 wt_1 = wt_2 = 0$
Osim	1.0	0.50	0.13	0.57	0.47
Ksim	1.0	0.28	0.26	0.28	0.28
RD	0.0	2002.97	5518.22	1841.13	1502.19
WRD	0.0	6.87	8.80	6.34	7.91

Table 3 shows the top-5 authors for each of the four T-Rank settings, in comparison to static PageRank. In T-Rank1 weight is evenly distributed among wt_i where as in T-Rank2 all emphasis is given to the freshness of a target page (i.e., $wt_1=1$). The respective results are quite similar. Compared to static PageRank this increases the ranking of some authors with fairly good static authority but were not among the top-5 PageRank results, namely, Lorie, Astrahan, Price, Traiger, who were not that active in the *twi* but co-authored one paper in 1999 (a System R retrospective on the occasion of its 20th anniversary), thus "refreshing" their formerly high authority. This effect would not have occurred if our *twi* had been five years earlier. This is in contrast to the static rank-1 author, Codd, who is not among the top T-Rank results simply because his node in the graph is not updated anymore through new publications in the *twi* of interest.

In the parameter setting T-Rank3, all emphasis is on the freshness of the individual edge along which authority is "transferred" in the random walk. This way authors who obtained a citation from another major authority within the *twi* moved up in the ranking. This holds, for example, for Melton and Eisenberg, authors of the SQL-1999 book; albeit not cited that frequently (within the DBLP fragment that has citations listed), these authors obtained fresh citations by some key authorities.

Finally, in the T-Rank4 setting the emphasis is on the average freshness of a target node's incoming links. This prioritizes authors who have many fresh citations in the *twi* of interest. Thus, it is no surprise that Widom and Garcia-Molina appear as highly ranked authors, surely some of the most active and highly regarded database researchers of the recent past.

Table 3. The top-5 authors for the different T-Rank parameter settings

	PageRank	T-Rank1 $wt_1 = wt_2 = wt_3 = 1/3$	T-Rank2 $wt_1 = 1 wt_2 = wt_3 = 0$	T-Rank3 $wt_2 = 1 wt_1 = wt_3 = 0$	T-Rank4 $wt_3 = 1 wt_1 = wt_2 = 0$
1	E. F. Codd	Michael Stonebraker	Michael Stonebraker	Jim Melton	Jim Gray
2	Michael Stonebraker	Raymond A. Lorie	Raymond A. Lorie	Jim Gray	Jeffrey D. Ullman
3	Jim Gray	Morton M. Astrahan	Morton M. Astrahan	Andrew Eisenberg	Michael Stonebraker
4	Donald D. Chamberlin	Nathan Goodman	Irving L. Traiger	Michael Stonebraker	Hector Garcia-Molina
5	Raymond A. Lorie	Thomas G. Price	Thomas G. Price	Jeffrey D. Ullman	Jennifer Widom

4.2 The Web Experiment

In the second experiment, we applied T-Rank to a dataset resulting from a web crawl of the amazon.com portal. The crawl was focused to yield only product-specific pages, which provide customer reviews and references to similar products. In the end the crawl contained 206,865 pages along with the relevant links. The product-specific pages and the links between them form the graph structure that we used in the ranking computation. For every page the strings representing dates (e.g., of product reviews by customers) were extracted and the oldest of them was used as $TS_{creation}$ of the corresponding node while the remaining dates form the set $TS_{modifications}$. As for the links, we chose as $TS_{creation}$ the most recent of the two connected nodes' creation dates. The described mapping yielded no modifications for the edges, thus that the sets $TS_{modifications}$ are empty for edges. We ranked result lists for different queries as follows. First, all pages relevant to the query (i.e. pages containing the query terms) were identified,. Second, we eliminated duplicate product ids, that is if multiple results' pages refer to the same product, all but one were eliminated. Third, having identified the relevant pages, we produced ranked result lists using both PageRank and T-Rank. Table 4 and Table 5 depict the results (i.e. for better comprehension we refer to the product name and not to the URL) for the queries "Harry Potter" and "Michael Moore" as returned by PageRank and T-Rank.

Table 4. Top-5 for the query "Harry Potter" as returned by PageRank and T-Rank

	PageRank
1.	GirlWise: How to Be Confident, Capable, Cool, and in Control (Julia Devillers)
2.	Harry Potter Potion Of Drought Apothecary Kit (Delta Education)
3.	You and Your Adolescent Revised Edition : Parent's Guide for Ages 10-20, A (Laurence Steinberg)
4.	The Art of The Return of the King (The Lord of the Rings) (Gary Russel)
5.	The Return of the King Visual Companion: The Official Illustrated Movie Companion

	T-Rank
1.	Harry Potter and the Prisoner of Azkaban Color and Activity Book (Harry Potter)
2.	Harry Potter and the Goblet of Fire (Harry Potter) (J.K. Rowling)
3.	The City of Ember [UNABRIDGED] (Jeanne Duprau, Wendy Dillion)
4.	Harry Potter and the Order of the Phoenix (Book 5) (J.K. Rowling)
5.	Hellboy: The Art of the Movie (Guillermo Del Toro)

It is important to place and consider the results in the temporal context of the crawl – May 2004. As depicted in Table 4, the T-Rank method ranks a product which is related to the current Harry Potter movie to the top, whereas no comparable product appears in the Top-5 obtained from PageRank. In total the Top-5 returned by T-Rank contains three items, which can be easily verified as relevant to the query, whereas only one such element appears in the Top-5 according to PageRank. Similar results emanate from the second query "Michael Moore". None of the products returned in the PageRank's Top-5 is clearly relevant to the query. On the contrary, T-Rank's Top-5 list contains four products related to the popular author/director.

Table 5. Top-5 for the query "Michael Moore" as returned by PageRank and T-Rank

	PageRank
1.	Esquire's Things a Man Should Know About Style (Scott Omelianuk, Ted Allen)
2.	Everything About Me Is Fake . . . And I'm Perfect (Janice Dickinson)
3.	Against All Enemies: Inside America's War on Terror (Richard A. Clarke)
4.	Lies: And the Lying Liars Who Tell Them.A Fair and Balanced Look at the Right (Al Franken)
5.	Bushwhacked : Life in George W. Bush's America (Molly Ivins, Lou Dubose)
	T-Rank
1.	Lies: And the Lying Liars Who Tell Them.A Fair and Balanced Look at the Right (Al Franken)
2.	Dude, Where's My Country? (Michael Moore)
3.	Dude, Where's My Country? Audio Book (Michael Moore)
4.	Dude, Where's My Country? Lage Print (Michael Moore)
5.	Stupid White Men ...and Other Sorry Excuses for the State of the Nation! (Michael Moore)

For lack of space we cannot present full results in a systematic way. We merely mention that in all test cases the results showed that T-Rank brings up the truly important and timely results to high ranks.

5   Conclusion

The Web already spans more than a decade of intensive use. If we assume a potential Web archive its size and complexity would be vast. On the other hand such an archive would be a valuable corpus for queries with strong temporal flavor. The method developed in this paper aim to help users finding Web pages that are both important (for a given query or topic of interest) and new (regarding the user's time window of interest). By incorporating temporal annotations to nodes and edges of the Web graph and appropriately extending the Markov chain model for computing, now time-aware, authority scores, our method can focus either on time snapshots or on temporal intervals. We believe that time-awareness is a very important issue in the quest for more meaningful ranking models in web search. The initial results presented in this paper are encouraging and motivate us to continue our effort towards 1) large scale experiments on web data crawls that span long periods and involve blind testing techniques to acquire end user judgments, and 2) extensions of the T-Rank model to assess the emerging authority of pages and identify trends in the evolving web.

References

1. Digital Bibliography and Library Project, http://dblp.uni-trier.de/
2. Internet Archive, available at: http://www.archive.org
3. E. Amitay, D. Carmel, M. Hersovici, R. Lempel, A. Soffer, U. Weiss: *Temporal link analysis*. Technical Report, IBM Research Lab Haifa, 2002.
4. R. Baeza-Yates, F. Saint-Jean, C. Castillo: *Web Structure, Dynamics and Page Quality*. String Processing and Information Retrieval Workshop, 2002.

5. A. Borodin, G. O. Roberts, J. S. Rosenthal, P. Tsaparas: *Finding Authorities and Hubs From Link Structures on the World Wide Web.* Proceedings of the 10th International World Wide Web Conference, pp. 415-429, 2001.

6. A. Broder, R. Kumar, F. Maghoul, P. Raghavan, S. Rajagopalan, S. Stata, A. Tomkins, J. Wiener: *Graph structure in the web.* Proceedings of the 9th WWW conference, 2001.

7. R. Fagin, R. Kumar, D. Sivakumar: *Comparing top k lists.* SIAM J. Discrete Mathematics 17 (1), pp. 134-160, 2003.

8. D. Fetterly, M. Manasse, M. Najork, J. Wiener: *A Large-scale Study of the Evolution of Web Pages.* Software Practice and Experience 34, pp. 213-237, 2004.

9. T.H. Haveliwala: *Topic-Sensitive PageRank: A Context-Sensitive Ranking Algorithm for Web Search.* IEEE Trans. Knowl. Data Eng. 15(4): 784-796 (2003).

10. G. Jeh, Jennifer Widom: *Scaling Personalized Web Search.* Proceedings of the twelfth international conference on World Wide Web, pp. 271-279, 2003.

11. M. Kendall, J.D Gibbons: *Rank Correlation Methods.* Oxford Univ. Press, 1990.

12. Jon M. Kleinberg: *Authoritative sources in a hyperlinked environment.* Journal of the ACM 46 (5), pp. 604–632, 1999.

13. R. Kraft, E. Hastor, R. Stata: *TimeLinks: Exploring the link structure of the evolving Web.* Second Workshop on Algorithms and Models for the Web Graph (WAW), Budapest, 2003.

14. R. Kumar, P. Raghavan, S. Rajagopalan, D. Sivakumar, A. Tomkins, and Eli Upfal. *Stochastic models for the Web graph.* Proceedings of the 41th IEEE Symp. on Foundations of Computer Science, pp. 57-65, 2000.

15. L. Laura, S. Leonardi, G. Caldarelli and P. De Los Rios: *A Multi-Layer Model for the Web Graph,* 2nd International Workshop on Web Dynamics, 2002.

16. R. Lempel, S.Moran: *SALSA: The stochastic approach for link-structure analysis.* ACM Transactions on Information Systems, 19(2), pp. 131-160, 2001.

17. L. Page, S. Brin, R. Motwani, T. Winograd: *The PageRank Citation Ranking: Bringing Order to the Web.* Tech. Report, Stanford Univ., 1998.

18. S. Sizov, M. Theobald, S. Siersdorfer, G. Weikum, J. Graupmann, M. Biwer, P. Zimmer: *The BINGO! System for Information Portal Generation and Expert Web Search.* First Int. Conf. on Innovative Data Systems Research (CIDR), 2003.

19. G.-R. Xue, H.-J. Zeng, Z. Chen, W.-Y. Ma, H.-J. Zhang, C.-J. Lu: *Implicit link analysis for small web search.* ACM SIGIR Conference, 2003.

Links in Hierarchical Information Networks*

Nadav Eiron and Kevin S. McCurley

IBM Almaden Research Center

Abstract. We provide evidence that the inherent hierarchical structure of the web is closely related to the link structure. Moreover, we show that this relationship explains several important features of the web, including the locality and bidirectionality of hyperlinks, and the compressibility of the web graph. We describe how to construct data models of the web that capture both the hierarchical nature of the web as well as some crucial features of the link graph.

1 Introduction

One feature that seems to have been largely ignored in studies of the Web is the inherent hierarchy that is evident in the structure of URLs. For example, in the URL http://ibm.com/products/server/ we might expect to find product information about servers, and below this we might expect a layer containing information about the individual models. In addition to the hierarchical structure of file paths on a server, URLs also reflect another layer of hierarchy from the domain name system (DNS). The hierarchical structure of the web reflects a practice that is at least 2000 years old, namely that information is often organized in a hierarchical tree structure, with information at the upper levels of the tree being more general than the information at the bottom levels[1].

In his seminal work on complex systems, Simon [1] argued that all systems tend to organize themselves hierarchically. Moreover, he stated that:

> "If we make a chart of social interactions, of who talks to whom, the clusters of dense interaction in the chart will identify a rather well-defined hierarchic structure."

We believe that a similar phenomenon can be seen in the link structure of the World Wide Web, in which a large fraction of the hyperlinks between URLs tend to follow the hierarchical organization of information and social groups that administer the information. In particular, we shall provide evidence that hyperlinks tend to exhibit a "locality" that is correlated to the hierarchical structure of URLs, and that many features of the organization of information in the web are predictable from knowledge of the hierarchical structure.

---

* An extended version is available at http://www.mccurley.org/papers/entropy.pdf
[1] We follow the convention that trees have their leaves at the bottom.

S. Leonardi (Ed.): WAW 2004, LNCS 3243, pp. 143–155, 2004.

Our contributions are in three areas. First, we describe the results of some statistical analysis carried out on an extremely large sample from the World Wide Web. We then point out how previous models of the web fail to adequately predict some important characteristics of the web, including the correlation between hierarchical structure and link structure, a measure of entropy for links (as evidenced by compressibility of the web graph), the growth of individual web sites, and bidirectionality of links on the web. We then describe a new hierarchical approach to modeling the web that incorporates these characteristics. Our primary focus is on describing a class of models and how the observed structure is reflected in the model, rather than on precise mathematical analysis of a particular abstract model.

While the motivation for this work arises from the information structure of the World Wide Web, we expect this feature to appear in other information networks for which there is an obvious hierarchical information organization, e.g., scientific citations, legal citations, patent citations, etc. In the extended version of this paper we show how similar patterns can be found in the instant messaging patterns within corporations.

2   Previous Models of the Web

In recent years there has been an explosion of published literature on models for networked systems, including the World Wide Web, social networks, technological networks, and biological networks. For coverage of this we refer the reader to the survey by Newman [2]. Much of this work is in the spirit of Simon's work on complex systems; attempting to explain various features that are ubiquitous across very different kinds of systems. Examples include small world structure, degree distributions, and community structure. Beyond these generic characteristics that show up across many different classes of networks, there are other features that may be unique to a particular type of network such as the World Wide Web. Some of these are due to the directed nature of the Web, but others are specific to the structure of information that the Web represents.

Models of the web are generally defined as stochastic processes in which edges and nodes are added to the graph over time in order to simulate the evolution of the web (or any other network). Such models fall broadly into two categories. The first category is those that rely upon Price's concept of *cumulative advantage*, also sometimes referred to as *preferential attachment* or "the rich get richer". In this model, the probability of adding an edge with a given destination and/or source is dependent on the existing in or out degree of the node (usually in a linear fashion). The second class of models uses a notion of *evolving copying*, in which the destinations for edges from a node are copied as a set from an existing node chosen under some distribution [3].

In section 6 we will present a new paradigm for constructing models of information networks that incorporates their hierarchical structure. It is our hope that by breaking the web down into the component features of site size, hierarchical structure of information, and link structure, we will present a useful paradigm for future analysis that incorporates multiple features of the web. It should be noted that the hierarchical evolution of structure can be combined with previous techniques of cumulative advantage or copying.

A hierarchical model of the web was previously suggested by Laura et. al. [4]. In their model, every page that enters the graph is assigned with a constant number of abstract "regions" it belongs to, and is allowed to link only to vertices in the same region. This forces a degree of locality among the vertices of the graph, though the definition of regions is unspecified, and the model artificially controls connections between these regions. In our model, we use the explicit hierarchy implied in the structure of URLs to establish the regions, which reflects a social division by organization.

Another recent model that incorporates hierarchical structure was proposed in [5]. Their model is generated in a very regular fashion, by starting with a small graph of five nodes, and replicating it five times, and joining these replicas together, and recursing this procedure. The resulting graph is shown to exhibit a clustering coefficient that resembles many real networks. Another recent model that results in a hierarchical organization of nodes was proposed in [6]. In both cases the models are fairly simple, and are designed to produce some specific generic properties such as clustering coefficient and degree distributions.

3   Experimental Methodology

Our observations are based on examination of a large subset of the Web that has been gathered at IBM Almaden since 2002. At the time of our experiments, the crawl had discovered at least 5.6 billion URLs on over 48 million hosts. For our analysis of tree structure we used the complete set of URLs, and for our analysis of link structure we used the first billion crawled URLs. For some of our experiments, we sampled from among the crawled URLs in smaller proportion in order to keep the computations manageable. Our goal was to use as large a data set as possible in order to provide assurance that our observations are fairly comprehensive. Even with such a large data set, observations about the World Wide Web are complicated by the fact that the data set is constantly changing, and it is impossible to gather the entire web. The characteristics of the data set are also influenced by the crawl strategy used. The algorithm used by our crawler is fairly standard, and is well approximated by a breadth first order.

More than 40% of the URLs discovered in our crawl contain a ? character in them, and we excluded these from our study altogether because these URLs often represent relational data organization rather than hierarchical information organization. As the web continues to grow, we expect this feature to become increasingly important and future models should address their influence. A significant fraction of the sites and pages from the crawl were pornographic, and these sites are well known for unusual link patterns designed to manipulate search engine rankings. For this reason we used a simple classification scheme to remove these from our experimental set. We also restricted ourselves to HTML content, as other data types represented leaf nodes without outlinks.

4   The Web Forest

At a coarse level of granularity, we can think of the web as a collection of hosts that grow more or less independently of each other. The number of URLs per host was previously studied by Adamic and Huberman [7], who hypothesized that the growth of an individual web site is modeled as a multiplicative process, resulting in a lognormal distribution. This provides a model of the web as a mixture multiplicative processes, leading to a power law tail for the distribution. Indeed, our large data set seems to confirm this conclusion, as it exhibits a power law distribution in the tail (though there is some variation with a hump in the middle). There are several subtleties that underly the use of multiplicative models (see [8]), but space limitations forbid us from further discussion on this point.

4.1   Tree Shapes

It is also of interest to investigate the distribution of the number of directories per site. Based on observations from 60 million URLs on 1.3 million sites, it was previously observed in [9] that the size of directory subtrees at a given depth appears to follow a power law, which is consistent with our observations on a much larger data set. We also observed that the number of directories per site behaves more like a pure power law.

One might wonder how the shapes of directory trees of web servers are distributed, and how the URLs on a web server are distributed among the directories. For this purpose, we sorted the static URLs in our set of 5.6 billion URLs by directory order. Then for each directory we computed the number of URLs that correspond to the files in that directory, the number of subdirectories, and the depth of the directory. In our data set we found that the number of URLs per directory and the number of subdirectories per directory exhibit a power law distribution. We estimate that the probability of finding more than n URLs in a subdirectory is approximately $c/n^{1.20}$ for large n, and the probability of finding more than d subdirectories of a directory is approximately $c/d^{1.43}$.

One might wonder whether the process of creating directories and creating URLs within the directory are correlated to each other, either positively or negatively, i.e., whether the existence of many URLs in a non-leaf directory is correlated to whether the directory has many (or few) subdirectories. In order to test this hypothesis, we computed a Goodman-Kruskal Gamma statistic on fanouts and URL counts for a sample of non-leaf directories from our data set. Our computations suggest that they are only slightly concordant, so it is probably safe to model them as independent processes.

4.2   Growth of Trees

In order to understand the distribution of links in the hierarchy, we first need to understand the structure of directory trees for web sites. At least two models for the shape of random trees have been studied extensively, namely the class of random recursive trees and the class of plane-oriented random recursive trees (see [10]). Random recursive trees are built up by selecting a node uniformly at random and attaching a new child to it. This class of trees results in a fanout distribution where $\Pr(\text{degree} = k) \sim 2^{-k}$, and are thus unsuitable for describing the type of trees seen here. By contrast, the construction of plane oriented trees chooses a node with probability proportional to 1+degree, resulting in a fanout distribution where $\Pr(\text{degree} = k) \approx \frac{4}{(k+1)(k+2)(k+3)}$, or $\approx 4/k^3$ for large k. This therefore gives a power law for the fanout distribution, but with the wrong exponent. It is however a simple matter to modify the plane-oriented model to incorporate a mixture between cumulative advantage and uniform attachment in order to achieve a more realistic degree distribution for trees.

The web does not grow as a single tree however; it grows as a forest of more or less independent trees. Rather little has been written about models for random forests, but one model was studied in [11]. Among other things, they proved that their model the number of trees in a forest of size n nodes is asymptotically $\log n$. In our data set from the web, we found approximately 48 million trees (websites) in a forest of 417 million nodes (directories). For this reason alone, the model of [11] does not seem appropriate as a model of the web forest, as it would have predicted a much smaller number of websites.

Another potential problem is related to the fact that web sites tend to grow largely independent of each other, whereas in [11] the placement of new leaves in the forest is dependent on the structure of the entire existing forest. In reality, the particular size of one web site usually has no bearing on the size of another web site (excluding mirrors and hosts within a single domain). For this reason we believe it is natural to model the growth of the forest as a collection of independent trees.

Random recursive forests were also considered by Mitzenmacher [12] in the context of modeling file sizes. In his model, files are either created from scratch according to a fixed distribution, or else they are created by copying an existing file and modifying it. The trees are then used to model the evolution of files, and he used a constant probability of creating a new root at each step, resulting in many more trees in the forest than the uniform recursive model of [11]. We adopt a similar strategy in our model of web forest growth, by maintaining a constant probability of creating a new web site at each time step.

There are a number of interesting statistics that might be investigated concerning the growth of the web forest. One difference between the two classes of random trees is found in the number of leaves. For random recursive trees the expected number of leaves is asymptotically $1/2$ the number of nodes, whereas for plane-oriented random recursive trees the expected number of leaves is asymptotically $2/3$ of the number of nodes. In the case of our web sample, we found that for hosts with more than 10 directories, the average number of leaves was 60%, and across the entire web the number of leaves is 74%. Hence the presence of many small sites on the web with few directories contributes many of the leaves. Note also that the 60% figure agrees with our suggestion to interpolate between plane-oriented trees and uniform recursive trees.

5 Link Locality

It has been observed in various contexts that links on the web seem to exhibit various forms of "locality". We loosely use the term locality to mean that links tend to be correlated to pages that are "nearby" in some measure. The locality of links in the web is important for various applications, including the extraction of knowledge on unified topics and the construction of efficient data structures to represent the links. We shall consider the latter issue in Section 7.

Numerous types of locality measures have been studied previously, and we refer to Newman's survey [2, § IIIB]. While previous studies have validated the existence of locality among links in the Web, they have not shed much light on the *process* that creates the locality, nor do they take into account the purposes for which links are created. We believe that much of the locality of links can be explained by a very strong correlation between the process of creating links and that of growing the hierarchy of a web site. Specifically, links can be of two types: navigational links within a cohesive set of documents, and informational links that extend outside the corpus developed by the author. Navigational links can further be broken down into templatized links designed to give web pages a common look and feel, and informational links that facilitate exploration of a body of knowledge. We divide links into six distinct types: Self loops, Intra-directory links, Up and Down links (those that follow the directory hierarchy), Across links (all links within a host that are not of the other types), and External links that go outside of the site. The second column of Table 1 shows the distribution of links into the various types, based on a sample of links from our entire corpus. This data clearly shows that external links are relatively rare, particularly when considering the fact that picking end points for links randomly by almost any strategy would result with almost all links being external. Note that when we limit ourselves to links for which we have crawled both ends, the fraction of external links is even smaller. This is partly because "broken" links are more common among external links, and partly because of our crawling strategy.

The discrepancy between the number of down and up links is perhaps surprising at first, but reflects several factors. First, many sites have every page

Table 1. Distribution by type for a sample of links. Shown are a sample of links where both source and destination are static URLs, and the subset where both ends were crawled. In the final column we tabulate the number of bidirectional links. Self loops (which were not included in the sample) account for roughly 0.9% of the links

Type of link	Static links	Both ends crawled	Bidirectional
Intra-directory	32.3%	41.1%	80.3%
Up	9.0%	11.2%	4.5%
Down	5.7%	3.9%	4.5%
Across directories	18.4%	18.7%	10.0%
External to host	33.6%	25.0%	0.7%
Total	5.1 billion	534893	156859

Table 2. Distribution of intra-host links in our test corpus and in a randomly generated graph on a sample of sites. Random assignment produces a distinctly different distribution of link types

Type of link	Crawled links	Random links
Internal	48.6%	32%
Up	13.6%	6%
Down	8.6%	5%
Across	22.7%	57%

equipped with a link to the top of the site (i.e., the starting point of the site), but downward links often target a single "entry page" in a directory [13]. Second, resource limitations on crawling and author-imposed restrictions on crawling via a `robots.txt` file will result in some down links being discovered without crawling the lower level pages to discover up links.

Another point one may consider when examining the distribution of links of the various types is the influence of normalizing the distribution by the number of possible targets of the various types. For example, in a random sample of approximately 100,000 web sites, we found that approximately 92% of the URLs appear at the leaves of the directory tree. Clearly, leaves cannot have outgoing "down" links.

How much does the tree structure dictate the distribution we see? To answer this question we picked a random sample of roughly 100,000 sites, and for each page, generated outlinks to other pages from the same site uniformly at random. We generated the same *number* of outlinks as the pages originally had. We compare this to the distribution of types of outlinks in general, normalized to exclude self-loops and external links, in Table 2. The data clearly shows a significantly higher number of links that follow the hierarchy (intra-directory, up and down links) in the real data, compared to what a random selection of targets will generate. This shows that the creation of links is highly correlated with the hierarchical structure of a web site.

Another measure of locality that bears some relationship to the hierarchical structure is the measure of directory distance. This distance is calculated by considering the directory structure implicitly exposed in a URL as a tree, and measuring the tree traversal distance between the directories (e.g., the number of directories between slashes that must be removed and appended to get from one URL to the other). For external links we add 1 for a change of hostname. We found that the probability of a link covering a distance d decreases exponentially in d, in spite of the fact that the number of eligible targets for a link initially *increases* with the distance.

5.1   Hyperlink Bidirectionality

In order for two web pages to share links to each other, the authors must at least know of their existence. Thus if the pages are created at different times, the page created first must either be created with a "broken" link, or else it is later modified to include a link to the page created later. In the case when pages are created by different authors, either they must cooperate to create their shared links, or else one page must be modified after creation of the other. This may explain why many bidirectional links appear between pages that are authored by the same person at the same time.

For these experiments, we used roughly the first 600 million pages from the crawl. In order to examine the existence of bidirectional links in our corpus, we randomly sampled 1/64th of the URLs, recording the links between pairs of pages that had been crawled, and noting when there were links going in each direction between the pair of pages. The results, broken down by link type, are shown in Table 1. From this data we can draw several conclusions. First, bidirectional links are far more frequent than previous models would have predicted. Second, it is evident that the vast majority of bidirectional links occur in the same directory, and probably arise from simultaneous creation by the same author. Bidirectional links between pages on dissimilar sites are extremely rare and probably indicates a high degree of cooperation or at least recognition between the authors of the two pages.

6   Hierarchy in Models for the Web

We believe that the approach to modeling the web should incorporate the social process of authorship, and the nature of social relationships within increasingly larger groups. The hierarchical structure of the social groups of authors of web information follows very closely the development of other social phenomenon as described by Simon [1]. In addition to this social hierarchy, web information often has a topical hierarchy associated with it that is often recognizable from the URL hierarchy.

We propose a model in which the web grows in two different (but related) ways. First, new hostnames get added to the web, and second, new URLs get added to existing hosts. We treat these processes separately, by evolving two graph structures for the forest directory structure and the hyperlinks. Sites themselves grow in a hierarchical fashion, with a site starting as a single URL, and growing into a tree. There are many variations on the procedure that we describe, and we defer discussion of these until after we describe the basic model.

We first describe how the forest structure is built. At each step in time a new URL is added to the Web. With probability ϵ, this URL is added as a new tree (i.e., a new site), containing a single URL. With probability η we create a new directory on an existing site to put the URL into. With probability γ we pick an existing leaf directory (a directory that has no sub-directories) and add the new URL to it. Finally, with probability $1 - \gamma - \epsilon - \eta$, we pick an existing non-leaf directory and add the new URL to it. In the case where a new directory is to be created, we pick the parent directory uniformly at random with probability c_f, and with probability $1 - c_f$, in proportion to the current fanout of the directory. When adding a URL to an existing directory, we pick a directory uniformly at random with probability c_s, and with probability $1 - c_s$ with proportion to the number of URLs in the directory.

We now describe how the links are created. At the time that we create a URL, we create a single inlink to the newly created page (this makes the resulting graph connected). If the URL is created on a new site, the source for the inlink is chosen uniformly at random from all URLs in the graph. If it is created in an existing site, we pick a URL uniformly at random from the set of URLs in the directory where the new URL is attached and the directory immediately above it.

We now have to say how to create links from each newly created URL. We hypothesize the existence of five parameters that are preserved as the graph grows, namely the probabilities of a link being internal, up, down, across, or external. For each type of link t we have a fixed probability p_t that remains constant as the graph grows, and $\sum_t p_t = 1$. For each type of link we also have a fixed probability b_t that the link will be bidirectional. In assigning links from a page, we first decide the number of links in accordance with a hypothesized distribution on the outdegrees from pages. We expect that this distribution has a power law tail, but the small values are unimodal and dictated by the common conventions on page design (in our simulation we use the observed outdegree distribution averaged over all sites). For each created link we assign it a type t with probability p_t. We pick the target for the link from among the eligible URLs with a mix of uniform and preferential attachment, namely with probably δ we choose one uniformly at random, and with probably $1 - \delta$ we pick one with probability that is proportional to its existing indegree. If there are no eligible URLs to create a link to, then we simply omit the link (for example, in the case of attempting to create a down link from a URL at a leaf directory). If we create a link, then we create a backlink from that link with probability b_t.

The mix of uniform and preferential attachment for inlinks is designed to guarantee the power law distribution for indegree. There are endless variations on this model, including the incorporation of copying, a preference for linking to URLs that are a short distance away, preferences for linking to URLs that are at a given level of the directory tree, etc. The purpose of our exposition here is to propose a simple model that satisfies the hierarchical requirement mentioned previously.

7   Link Compression and Entropy

It has been observed by several authors that the link graph is highly compressible, culminating in the recent work of Boldi and Vigna [14] on encodings that use only 3 bits per link. If the links were *random* then of course this would not be possible, as an easy probabilistic argument says that at least 28 bits would be required to store a single link from each page, and this number would grow as $\log(N)$ for a graph with N nodes. One possible source of redundancy in the link structure may be attributed to the power law distribution of indegrees. However, it was observed by Adler and Mitzenmacher [15] that a simple Huffman encoding scheme that exploits only this redundancy for compression of the web graph would still require $\Omega(\log(N))$ bits to represent an edge in an N-node link graph. This suggests that there are other sources of redundancy in the link graph that allow for such high levels of compression.

In fact, the hierarchical locality for links that we have observed is closely related to why such good compression schemes for the web graph are achievable. One method is to sort the URLs lexicographically, and encode a link from one URL to another by the difference between their positions in the list. This delta encoding is small precisely because the URLs of source and destination often agree on a long prefix of the strings, and are therefore close together in a lexicographic sort order. Since lexicographic order of URLs is a good approximation of directory order, the compressibility of the link graph is closely related to the locality of links in the hierarchical structure. This observation that locality is the source of compressibility of the web graph was also made in [15].

To further explore the link between hierarchical structure and compression, we wish to examine how well various web models explain link compressibility. Rather than considering the various compression schemes devised (which are mostly based on the textual structure of the URLs, and are designed to facilitate efficient implementation), we concentrate on the information-theoretical measure of the *entropy* of the link graph. We choose to focus on the following entropy measure, which we call *isolated destination entropy*. We define the probability distribution whose entropy we measure as follows: First, the evolutionary model is used to grow a graph to a given number of nodes N. Then, we consider the distribution of the destination URLs that are linked to from each URL (where the distribution is over the set of all nodes in the graph). In other words, for each source URL v, we consider the distribution of the random variable D_v whose values are the destinations of links originating at v.

Our motivation in picking the entropy measurement to be based on a static snapshot of the graph, rather than considering the entropy of the selection process employed by the various evolutionary model, is to mimic the conditions faced by a compression algorithm for a web. When compressing the web, a compression algorithm typically has no knowledge of the order in which URLs and links were created. Furthermore, we would like a measure of compressibility that is independent of the evolutionary model used, to allow for an apples-to-apples comparison of the various models. The isolated destination entropy is a lower bound on the compression that may be achieved by compression schemes that encode each destination independently of other destinations. It obviously also dictates a lower bound for more sophisticated methods.

This measure captures the redundancy in information that is present because outlinks from a given page are typically made to pages in the close proximity to the source page. However, this does not capture the more global phenomenon that makes pages that are close to each other in the hierarchy have links similar to each other. In fact, it does not even directly exploit the dependency between different pages linked to from the same page. The effects of this phenomenon are part of the explanation for the improvements over the Link2 scheme in [16] achieved by schemes, such as the Link3 scheme in [16], that use delta encodings between outlink lists of multiple pages.

Unfortunately, because of the complexity of the models and the fact that we are measuring entropy on snapshots of the graph, we are unable to analytically compute the isolated destination entropy. Instead, we provide empirical measurements for various models. To measure the isolated destination entropy we use each model to generate 225 random graphs, each containing a million nodes. Where applicable, we use the same arbitrary set of URLs (and hierarchical structure) for all graphs generated by a model, and only allow the link generation process to be driven by a pseudo-random number generator. We then sample a fraction of the nodes in all the graphs, and empirically estimate their average isolated destination entropy by calculating the entropy of the empirical distribution of outlinks from a node. We express our entropy measurement in bits per link, as is customary in works that describe compression schemes for the web graph [16, 15]. When comparing these results to the theoretical maximum entropy, one must note that because of the relatively small sample that we use relative to the domain of the random variable D_v, the upper bound on the entropy is much lower than the usual $\log N$ for a graph with N nodes. Instead, if the average outdegree is d, and we generate m distinct graph, $\log(md)$ is an upper bound on the empirical isolated destination entropy we can expect. This is because, on average, only md outlinks from any given node will be encountered in the sample.

In what follows we compare three models. First is PAModel which is a preferential attachment model based on [17, 18]. In this model, the destination for outlinks is chosen by a mixture of preferential attachment and a uniform distribution over all previously cerated URLs. The second is called AMModel, which is similar to that of Kumar et al. [19]. Following Adler and Mitzenmacher's choice of parameters for this model labeled G_4 [15], we set the parameters for this model to copy links from zero to four previous nodes, where each link is copied with probability 0.5, and either one or two additional links are then added with the destination chosen uniformly at random. We compare these two models with our hierarchical model as described in Section 6. Both PAModel and the hierarchical model require outdegrees to be drawn from a power law distribution. Rather than using a pure power law distribution, we use a sample of outdegrees from our web crawl to determine the "head" of the outdegree distribution in these models, with the tail being determined by a power law. This distribution has a mean of about 27 outlinks. The Hierarchical model exhibits a slightly lower average outdegree in practice, because some outlinks may not be feasible (e.g., uplinks from top level URLs, etc.). The results of our experiments are summarized in Table 3. The results clearly point out that the destinations for outlinks in our hierarchical web model are far less random than those generated by the previous models we compare against. The results also demonstrate that graphs generated by an evolutionary copying model tend to have a less random structure than graphs where link destinations are chosen through a preferential attachment process. This suggests that incorporating copying into the hierarchi-

Table 3. Empirical measurements of isolated destination entropy on graphs generated by three models. Measurements are based on sampling from 225 graphs for each model, of size one million nodes each

Model	Empirical Entropy	Max. Entropy
PAModel	11.72	12.56
AMModel	9.6	12.05
Hierarchical	8.08	12.49

cal model may reduce the uncertainty in link creation even further, and yield an even more realistic model, as far as the measure of compressibility of the link graph is concerned.

8 Conclusions

In this work we have examined the interaction between two evolutionary processes that shape the web: the growth of hierarchical structures as reflected in URLs, and the creation of hyperlinks on the web. We have demonstrated that at least two features of the actual web graph, namely bidirectionality of links and compressibility links, are directly related to the hierarchical structure.

We have proposed a framework for models that incorporates an evolutionary process for both the hierarchical structure and the hyperlink graph. The model is further motivated by how web sites evolve, from the general to the specific. We believe that the hierarchical structure of the web will provide a mechanism for better understanding of the web, and hopefully lead to more effective algorithms for information retrieval, mining, and organization tasks.

References

1. Simon, H.A.: The Sciences of the Artifical. 3rd edn. MIT Press, Cambridge, MA (1981)
2. Newman, M.E.J.: The structure and function of complex networks. SIAM Review **45** (2003) 167–256
3. Kumar, R., Raghavan, P., Rajagopalan, S., Sivakumar, D.: Stochastic models for the Web graph. In: Proc. of the 41st IEEE Symposium on Foundations of Comp. Sci. (2000) 57–65
4. Laura, L., Leonardi, S., Caldarelli, G., Rios, P.D.L.: A multi-layer model for the web graph. In: 2nd International Workshop on Web Dynamics, Honolulu (2002)
5. Ravasz, E., Barabási, A.L.: Hierarchical organization in complex networks. Phys. Rev. E **67** (2003)
6. Chakrabarti, D., Zhan, Y., Faloutsos, C.: R-MAT: A recursive model for graph mining. In: Proc. SIAM Int. Conf. on Data Mining. (2004)
7. Huberman, B.A., Adamic, L.A.: Evolutionary dynamics of the world wide web. Technical report, XEROX PARC (1999)
8. Mitzenmacher, M.: A brief history of generative models for power law and lognormal distributions. Internet Mathematics **1** (2003) to appear.

9. Dill, S., Kumar, R., McCurley, K.S., Rajagopalan, S., Sivakumar, D., Tomkins, A.: Self-similarity in the web. ACM Transactions on Internet Technology **2** (2002) 205–223

10. Smythe, R.T., Mahmoud, H.M.: A survey of recursive trees. Theoretical Probability and Mathematical Statistics **51** (1995) 1–27 Translation from *Theorya Imovirnosty ta Matemika Statystika*, volume 51, pp. 1–29, 1994.

11. Balińska, K.T., Quintas, L.V., Szymański, J.: Random recursive forests. Random Structures and Algorithms **5** (1994) 3–12

12. Mitzenmacher, M.: Dynamic models for file sizes and double pareto distributions. Internet Mathematics (2004)

13. Eiron, N., McCurley, K.S.: Untangling compound documents in the web (2003) Proc. ACM Conf. on Hypertext and Hypermedia.

14. Boldi, P., Vigna, S.: The webgraph framework I: Compression techniques. In: Proc. Int. WWW Conf., New York (2004)

15. Adler, M., Mitzenmacher, M.: Towards compressing web graphs. Technical report, Harvard University Computer Science Dept. (2001) Short version in Data Compression Conference, 2001.

16. Randall, K.H., Stata, R., Wickremesinghe, R.G., Wiener, J.L.: The link database: Fast access to graphs of the Web. In: Proceedings of the 2002 Data Compression Conference (DCC). (2002) 122–131

17. Levene, M., Fenner, T., Loizou, G., Wheeldon, R.: A stochastic model for the evolution of the web. Computer Networks **39** (2002) 277–287

18. Pennock, D.M., Flake, G.W., Lawrence, S., Glover, E.J., Giles, C.L.: Winners don't take all: Characterizing the competition for links on the web. PNAS (2002) 5207–5211

19. Kumar, R., Raghavan, P., Rajagopalan, S., Tomkins, A.: Extracting large-scale knowledge bases from the Web. In Atkinson, M.P., Orlowska, M.E., Valduriez, P., Zdonik, S.B., Brodie, M.L., eds.: Proc. 25th VLDB, Edinburgh, Scotland, Morgan Kaufmann (1999) 639–650

Crawling the Infinite Web:
Five Levels Are Enough

Ricardo Baeza-Yates and Carlos Castillo

Center for Web Research, DCC,
Universidad de Chile
{rbaeza,ccastillo}@dcc.uchile.cl

Abstract. A large amount of publicly available Web pages are generated dynamically upon request, and contain links to other dynamically
generated pages. This usually produces Web sites which can create arbi-
trarily many pages. In this article, several probabilistic models for browsing "infinite" Web sites are proposed and studied. We use these models
to estimate how deep a crawler must go to download a significant portion
of the Web site content that is actually visited. The proposed models are
validated against real data on page views in several Web sites, showing
that, in both theory and practice, a crawler needs to download just a few
levels, no more than 3 to 5 "clicks" away from the start page, to reach
90% of the pages that users actually visit.

1 Introduction

Most studies about the web refer only to the "publicly indexable portion", excluding a portion of the web that has been called "the hidden web" [1] and is
characterized as all the pages that normal users could eventually access, but
automated agents such as the crawlers used by search engines do not. Certain
pages are not indexable because they require special authorization. Others are
dynamic pages, generated after the request has been made. Many dynamic
pages are indexable, as the parameters for creating them can be found by following links. This is the case of, e.g. typical product catalogs in Web stores, in
which there are links to navigate the catalog without the user having to pose a
query.

The amount of information in the Web is certainly finite, but when a dynamic
page leads to another dynamic page, *the number of pages can be potentially
infinite*. Take, for example, a dynamic page which implements a calendar; you
can always click on "next month" and from some point over there will be no more
data items in the calendar; humans can be reasonably sure that it is very unlikely
to find events scheduled 50 years in advance, but a crawler can not. There are
many more examples of "crawler traps" that involve loops and/or near-duplicates
(which can be detected afterwards, but we want to avoid downloading them).

In this work, we deal with the problem of capturing a relevant portion of
the *dynamically generated content with known parameters*, while avoiding the

S. Leonardi (Ed.): WAW 2004, LNCS 3243, pp. 156–167, 2004.
© Springer-Verlag Berlin Heidelberg 2004

download of too many pages. We are interested in knowing if a user will ever see a dynamically generated page. If the probability is too low, would a search engine like to retrieve that page? Clearly, from the Web site point of view the answer is yes, but perhaps from the search engine's point of view, the answer is no. In that case, our results are even more relevant. The answer in the case of the user's point of view is not clear a priori, as will depend on the result.

The main contributions of this paper are the models we propose for random surfing inside a Web site when the number of pages is **unbounded**. To do that, we take the tree induced by the Web graph of a site, and study it by levels. We analyze these models, focusing on the question of how "deep" users go inside a Web site and we validate these models using actual data from Web sites and link analysis with Pagerank. Our results help to decide when a crawler should stop, and to evaluate how much and how important are the non-crawled pages.

The next section outlines prior work on this topic, and the rest of this paper is organized as follows: in section 3, three models of random surfing in dynamic Web sites are presented and analyzed; in section 4, these models are compared with actual data from the access log of several Web sites. Section 5 concludes with some final remarks and recommendations for practical web crawler implementations.

2   Previous Work

Crawlers are an important component of Web search engines, and as such, their internals are kept as business secrets. Recent descriptions of Web crawlers include: Mercator [2], Salticus [3], WIRE [4], a parallel crawler [5] and the general crawler architecture described by Chakrabarti [6].

Models of random surfers as the one studied by Diligenti et al. [7] have been used for page ranking using the Pagerank algorithm [8], and for sampling the web [9]. Other studies about Web crawling have focused in crawling policies to capture high-quality pages [10] or to keep the search engine's copy of the Web up-to-date [11]. Link analysis on the Web is currently a very active research topic; for a concise summary of techniques, see a survey by Henzinger [12].

Log file analysis has a number of restrictions arising from the implementation of HTTP, specially caching and proxies, as noted by Haigh and Megarity [13]. *Caching* implies that re-visiting a page is not always recorded, and re-visiting pages is a common action, and can account for more than 50% of the activity of users, when measuring it directly in the browser [14]. *Proxies* implies that several users can be accessing a Web site from the same IP address. To process log file data, careful data preparation must be done [15], including the detection of sessions from automated agents [16].

The visits to a Web site have been modeled as a sequence of decisions by Huberman *et. al* [17, 18]; they obtain a model for the number of clicks that follows a Zipf's law. Levene et al. [19] proposed to use an absorbing state to represent the user leaving the Web site, and analyzed the lengths of user sessions when the probability of following a link increases with session length. Lukose and Huberman [20] also present an analysis of the Markov chain model of a user clicking through a Web site, and focus in designing an algorithm for automatic browsing, which is also the topic of a recent work by Liu et al. [21].

3   Random Surfer Models for a Web Site with Infinite Number of Pages

We will consider a Web site as a set of pages under the same host name, and a *user session* as a finite sequence of page views in this Web site. The starting point of a user session does not need to be the page located at the root directory of the server, as some users may enter to the Web site following a link to an internal page.

The *page depth* of a page in a session is the shortest path from the start page through the pages seen during a session. This is not only a function of the Web site structure, this is the perceived depth during a particular session. The *session depth* is the maximum depth of a page in a session.

For random surfing, we can model each page as a state in a system, and each hyperlink as a possible transition; or we can use a simpler model in which we collapse multiple pages at the same level as a single node, as shown in Fig. 1 (left and center). That is, the Web site graph is collapsed to a sequential list.

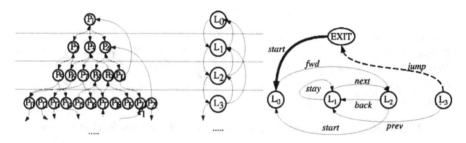

Fig. 1. Left: a Web site modeled as a tree. Center: the Web site modeled as a sequence of levels. Right: Representation of the different actions of the random surfer

At each step of the walk, the surfer can perform one of the following actions, which we consider as atomic: go to the next level (action *next*), go back to the previous level (action *back*), stay in the same level (action *stay*), go to a different previous level (action *prev*), go to a different higher level (action *fwd*), go to the start page (action *start*) or jump outside the Web site (action *jump*). For action *jump* we add an extra node EXIT to signal the end of a user session (closing the browser, or going to a different Web site) as shown in Fig. 1 (right). Regarding this Web site, after leaving, users have only one option: start again in a page with depth 0 (action *start*).

As this node EXIT has a single out-going link with probability 1, it does not affect the results for the other nodes if we remove the node EXIT and change this by transitions going to the start level L_0 . Another way to understand it is that as this process has no memory, going back to the start page or starting a new session are equivalent, so actions *jump* and *start* are indistinguishable in terms of the resulting probability distribution for the other nodes. The set of atomic actions is $\mathcal{A} = \{next, start/jump, back, stay, prev, fwd\}$.

The probability of an action at level ℓ is $Pr(action|\ell)$. As they are probabilities $\sum_{action \in \mathcal{A}} Pr(action|\ell) = 1$. The probability distribution of being at a level at a given time is the vector $\mathbf{x}(t) = (x_0, x_1, \dots)$. When there exists a limit, we will call this $\lim_{t \to \infty} \mathbf{x}(t) = \mathbf{x}$.

In this article, we study three models with $Pr(next|\ell) = q \; \forall \ell$, i.e.: the probability of advancing to the next level is constant for all levels. Our purpose is to predict how far will a real user go into a dynamically generated Web site. If we know that, e.g.: $x_0 + x_1 + x_2 \geq 90\%$, then the crawler could decide to crawl just those three levels. The models we analyze were chosen to be as simple and intuitive as possible, though without sacrificing correctness. We seek more than just fitting the distribution of user clicks, we want to *understand and explain user behavior in terms of simple operations*.

3.1   Model A: Back One Level at a Time

In this model, with probability q the user will advance deeper, and with probability $1 - q$ the user will go back one level, as shown in Fig. 2.

$Pr(next|\ell) = q,$
$Pr(back|\ell) = 1 - q$ for $\ell \geq 1,$
$Pr(stay|\ell) = 1 - q$ for $\ell = 0,$
$Pr(start, jump|\ell) = 0$ and
$Pr(prev|\ell) = Pr(fwd|\ell) = 0$.

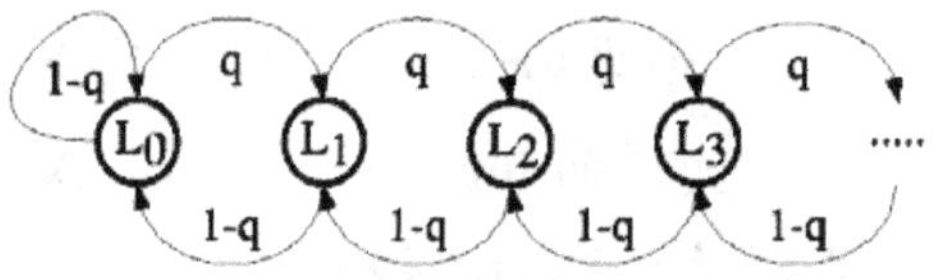

Fig. 2. Model A, the user can go forward or backward one level at a time

A stable state $\mathbf{x}$ is characterized by $\sum_{i \geq 0} x_i = 1$ and:

$$x_i = qx_{i-1} + (1 - q)x_{i+1} \quad (\forall i \geq 1),$$
$$x_0 = (1 - q)x_0 + (1 - q)x_1 .$$

The solution to this recurrence is: $x_i = x_0 \left(\frac{q}{1-q}\right)^i \quad (\forall i \geq 1)$.

If $q \geq 1/2$ then we have the solution $x_i = 0$, and $x_\infty = 1$ (that is, we have an absorbing state); which in our framework means that no depth can ensure that a certain proportion of pages have been visited by the users. When $q < 1/2$ and we impose the normalization constraint, we have a geometric distribution:

$$x_i = \left(\frac{1 - 2q}{1 - q}\right) \left(\frac{q}{1 - q}\right)^i .$$

The cumulative probability of levels $0 \ldots k$ is:

$$\sum_{i=0}^{k} x_i = 1 - \left(\frac{q}{1-q}\right)^{k+1} .$$

3.2   Model B: Back to the First Level

In this model, the user will go back to the start page of the session with probability $1 - q$. This is shown in Fig. 3.

$Pr(next|\ell) = q,$
$Pr(back|\ell) = 1 - q$ if $\ell = 1$, 0 otherwise,
$Pr(stay|\ell) = 1 - q$ for $\ell = 0$,
$Pr(start, jump|\ell) = 1 - q$ for $\ell \geq 2$ and
$Pr(prev|\ell) = Pr(fwd|\ell) = 0$.

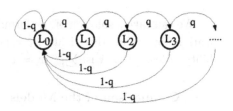

Fig. 3. Model B, the user can go forward one level at a time, or she/he can go back to the first level either by going to the start page, or by starting a new session

A stable state $\mathbf{x}$ is characterized by $\sum_{i \geq 0} x_i = 1$ and:
$$x_0 = (1 - q) \sum_{i \geq 0} x_i = (1 - q) ,$$
$$x_i = q x_{i-1} \quad (\forall i \geq 1) .$$

As we have $q < 1$ we have another geometric distribution: $x_i = (1-q)q^i$. The cumulative probability of levels $0 \ldots k$ is: $\sum_{i=0}^{k} x_i = 1 - q^{k+1}$.

Note that the cumulative distribution obtained with model A ("back one level") using parameter q_A, and model B ("back to home") using parameter q_B are equivalent if: $q_A = \frac{q_B}{1+q_B}$. So, as the distribution of session depths is equal, except for a transformation in the parameter q, we will consider only model B for charting and fitting the distributions.

3.3   Model C: Back to Any Previous Level

In this model, the user can either discover a new level with probability q, or go back to a previous visited level with probability $1 - q$. If he decides to go back to a previously seen level, he will choose uniformly from he set of visited levels (including the current one), as shown in Fig. 4.

A stable state $\mathbf{x}$ is characterized by $\sum_{i \geq 0} x_i = 1$ and:

$$x_0 = (1 - q) \sum_{k \geq 0} \frac{x_k}{k+1} ,$$

$$x_i = q x_{i-1} + (1 - q) \sum_{k \geq i} \frac{x_k}{k+1} \quad (\forall i > 1) .$$

$Pr(next|\ell) = q$,
$Pr(back|\ell) = 1 - q/(\ell + 1), \ell \geq 1$,
$Pr(stay|\ell) = 1 - q/(\ell + 1)$,
$Pr(start, jump|\ell) = 1 - q/(\ell + 1)$,
$\quad \ell \geq 2$,
$Pr(prev|\ell) = 1 - q/(\ell + 1), \ell \geq 3$,
$Pr(fwd|\ell) = 0$.

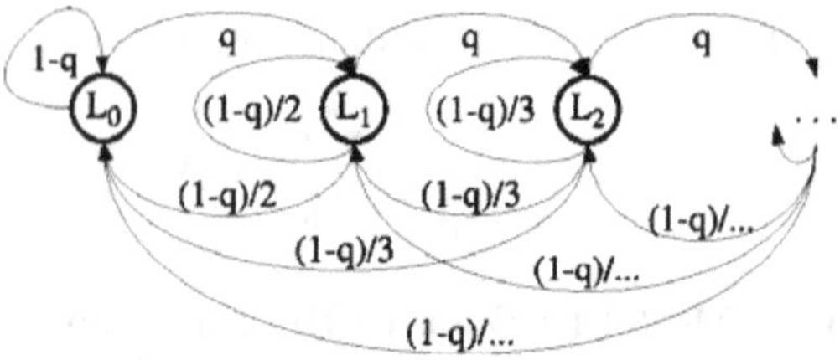

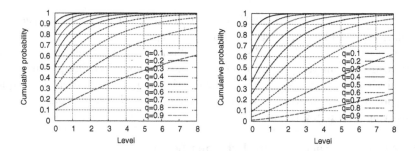

Fig. 4. Model C: the user can go forward one level at a time, and can go back to previous levels with uniform probability

We can take a solution of the form: $x_i = x_0 (i + 1) q^i$. Imposing the normalization constraint, this yields: $x_i = (1 - q)^2 (i + 1) q^i$. The cumulative probability of levels $0 \ldots k$ is: $\sum_{i=0}^{k} x_i = 1 - (2 + k - (k + 1) q) q^{k+1}$.

3.4 Comparison of the Models

In terms of the cumulative probability of visiting the different levels, models A and B produce equivalent results except for a transformation of the parameters. Plotting the cumulative distributions for models B and C yields Fig. 5. We can see that if $q \leq 0.4$, then in these models there is no need for the crawler to go past depth 3 or 4 to capture more than 90% of the pages a random surfer will actually visit, and if q is larger, say, 0.6, then the crawler must go to depth 6 or 7 to capture this amount of page views.

Fig. 5. Cumulative probabilities for models B (left) and C (right)

4   Data from User Sessions in Web Sites

We studied real user sessions on 13 different Web sites in the US, Spain, Italy and Chile, including commercial, educational, non-governmental organizations and Web logs (sites in in which collaborative forums play a major role, also known as "Blogs"); characteristics of this sample, as well as the results of fitting models B and C to the data are summarized in Table 1.

Table 1. Characteristics of the studied Web sites and results of fitting the models. The number of user sessions does not reflect the relative traffic of the Web sites, as the data was obtained in different time periods. "Root entry" is the fraction of sessions starting in the home page

	Collection						Fit		
Code	Type	Country	Recorded sessions	Average page views	Root entry	Best model	q	Error
E1	Educational	Chile	5,500	2.26	84%	B	0.51	0.88%
E2	Educational	Spain	3,600	2.82	68%	B	0.51	2.29%
E3	Educational	US	71,300	3.10	42%	B	0.64	0.72%
C1	Commercial	Chile	12,500	2.85	38%	B	0.55	0.39%
C2	Commercial	Chile	9,600	2.09	32%	B	0.62	5.17%
R1	Reference	Chile	36,700	2.08	11%	B	0.54	2.96%
R2	Reference	Chile	14,000	2.72	22%	B	0.59	2.75%
O1	Organization	Italy	10,700	2.93	63%	C	0.35	2.27%
O2	Organization	US	4,500	2.50	1%	B	0.62	2.31%
OB1	Organization + Blog	Chile	10,000	3.73	31%	B	0.65	2.07%
OB2	Organization + Blog	Chile	2,000	5.58	84%	B	0.72	0.35%
B1	Blog	Chile	1,800	9.72	39%	C	0.79	0.88%
B2	Blog	Chile	3,800	10.39	21%	C	0.63	1.01%

We obtained access logs with anonymous IP addresses from these Web sites, and processed them to obtain user sessions, considering a session as a sequence of GET requests with the same User-Agent [22] and less than 30 minutes between requests [23]. We also processed the log files to discard hits to Web applications such as e-mail or content management systems, as they neither respond to the logic of page browsing, nor are usually accessible by Web crawlers. We expanded sessions with missing pages using the Referrer field of the requests, and considering all frames in a multi-frame page as a single page. Finally, we discarded sessions by Web robots [16] using known User-Agent fields and accesses to the /robots.txt file, and we discarded requests searching for buffer overflows or other software bugs.

As re-visits are not always recorded because of caching [14], data from log files *overestimates the depth at which users spent most of the time*. Fig. 6 shows the cumulative distribution of visits per page depth to Web sites. At least 80%-95% of the visits occur at depth ≤ 4, and about 50% of the sessions include only the start page. The average session length is 2 to 3 pages, but in the case of Web logs, sessions tend to be longer. This is reasonable as Web postings are very short so Blog users view several of them during one session.

We fitted the models to the data from Web sites, as shown in Table 1 and Fig. 8. In general, the curves produced by model B (and model A) are a better approximation to the user sessions than the distribution produced by model C, except for Blogs. The approximation is good for characterizing session depth, with error in general lower than 5%.

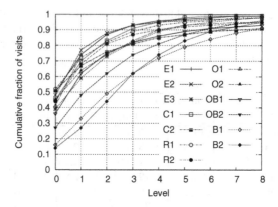

Fig. 6. Distribution of visits per level, from access logs of Web sites. E=educational, C=commercial, O=non-governmental organization, OB=Organization with on-line forum, B=Blog (Web log or on-line forum)

We also studied the empirical values for the distribution of the different actions at different levels in the Web site. We averaged this distribution across all the studied Web sites at different depths. The results are shown in Table 2, in which we consider all the Web sites except for Blogs.

Table 2. Average distribution of the different actions in user sessions, without considering Blogs. Transitions with values greater than 0.1 are shown in bold face

Level	Observations	Next	Start	Jump	Back	Stay	Prev	Fwd
0	247985	**0.457**	–	**0.527**	–	0.008	–	0.000
1	120482	**0.459**	–	**0.332**	**0.185**	0.017	–	0.000
2	70911	**0.462**	**0.111**	**0.235**	**0.171**	0.014	–	0.001
3	42311	**0.497**	0.065	**0.186**	**0.159**	0.017	0.069	0.001
4	27129	**0.514**	0.057	**0.157**	**0.171**	0.009	0.088	0.002
5	17544	**0.549**	0.048	**0.138**	**0.143**	0.009	**0.108**	0.002
6	10296	**0.555**	0.037	**0.133**	**0.155**	0.009	**0.106**	0.002
7	6326	**0.596**	0.033	**0.135**	**0.113**	0.006	**0.113**	0.002
8	4200	**0.637**	0.024	**0.104**	**0.127**	0.006	0.096	0.002
9	2782	**0.663**	0.015	**0.108**	**0.113**	0.006	0.089	0.002
10	2089	**0.662**	0.037	0.084	**0.120**	0.005	0.086	0.003

We can see in Table 2 that the actions *next*, *jump* and *back* are the more important ones, which is in favor of models A (back one level) and model B (back to start level). We also note that $Pr(next|\ell)$ doesn't vary too much, and lies between 0.45 and 0.6 . It increases as ℓ grows which is reasonable as a user that already have seen several pages is more likely to follow a link.

$Pr(jump|\ell)$ is higher than $Pr(back|\ell)$ for the first levels, and it is much higher than $Pr(start|\ell)$. About half of the user sessions involve only one page from the Web site. $Pr(start|\ell)$, $Pr(stay|\ell)$ and $Pr(fwd|\ell)$ are not very common actions.

5 Conclusions

The models and the empirical data presented lead us to the following characterization of user sessions: they can be modeled as a random surfer that either advances one level with probability q, or leaves the Web site with probability $1-q$. In general $q \approx 0.45-0.55$ for the first few levels, and then $q \approx 0.65-0.70$. This simplified model is good enough for representing the data for Web sites, but:

- We could also consider Model A (back one level at a time), which is equivalent in terms of cumulative probability per level, except for a change in the parameters. Based on the empirical data, we observe that users at first just leave the Web site while browsing (Model B), but after several clicks, they are more likely to go back one level (Model A).
- A more complex model could be derived from empirical data, particularly one that considers that q depends on ℓ . We considered that for our purposes, which are related to Web crawling, the simple model is good enough.
- Model C appears to be better for Blogs. A similar study to this one, focused only in the access logs of Blogs seems a reasonable thing to do since Blogs represent a growing portion of on-line pages.

In all cases, the models and the data show evidence of a distribution of visits which is strongly biased to the first few levels of the Web site. According to this distribution, more than 90% of the visits are closer than 4 to 5 clicks away from the entry page in most of the Web sites. In Blogs, we observed deeper user sessions, with 90% of the visits within 7 to 8 clicks away from the entry page.

In theory, as internal pages can be starting points, it could be concluded that Web crawlers must always download entire Web sites. In practice, this is not the case: if we consider the physical page depth in the directory hierarchy of a Web site, we observe that the distribution of surfing entry points per level rapidly decreases, so the overall number of pages to crawl is finite, as shown in Fig. 7 (left).

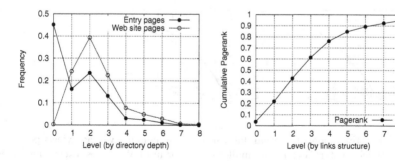

Fig. 7. Left: fraction of different Web pages seen at a given depth, and fraction of entry pages at the same depth in the studied Web sites, considering their directory structure. Right: cumulative Pagerank by page levels in a large sample of the Chilean Web

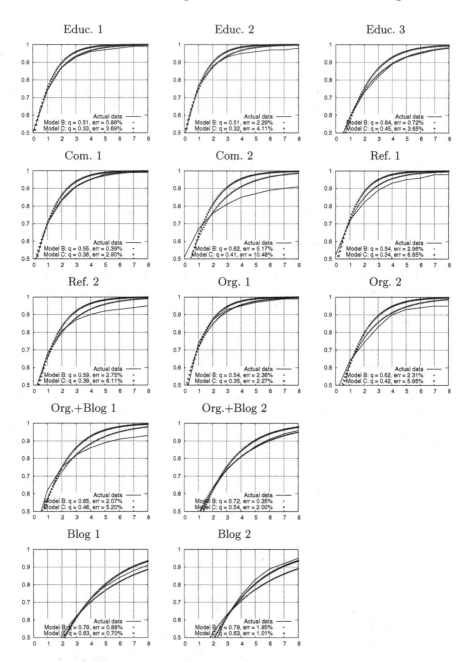

Fig. 8. Fit of the models to actual data, in terms of cumulative page views per level. Model B (back to start level), has smaller errors for most Web sites, except for Blogs. The asymptotic standard error for the fit of this model is 5% in the worst case, and consistently less than 3% for all the other cases. Note that we have zoomed into the upper portion of the graph, starting in 50% of cumulative page views

Link analysis, specifically Pagerank, provides more evidence for our conclusions. We asked, what fraction of the total Pagerank score is captured by the pages on the first ℓ levels of the Web sites? To answer this, we crawled a large portion of the Chilean Web (.cl) obtaining around 3 million pages on April 2004, using 150 thousand seed pages that found 53 thousand Web sites. Fig. 7 (right) shows the cumulative Pagerank score for this sample. Again, the first five levels capture 80% of the best pages. Note that the levels in this figure are obtained in terms of the global Web structure, considering internal and external links, not user sessions, as in the study by Najork and Wiener [10].

These models and observations could be used by a search engine, and we expect to do future work in this area. For instance, if the search engine's crawler performs a breadth-first crawling and can measure the ratio of new URLs from a Web site it is adding to its queue vs. seen URLs, then it should be able to infer how deep to crawl that specific Web site. The work we presented in this article provides a framework for that kind of adaptivity.

An interesting enhancement of the models shown here is to consider the contents of the pages to detect duplicates and near-duplicates. In our model, downloading a duplicate page should be equivalent to going back to the level at which we visited that page for the first time. A more detailed analysis could also consider the distribution of terms in Web pages and link text as the user browses through a Web site.

As the amount of on-line content that people, organizations and business are willing to publish grows, more Web sites will be built using Web pages that are dynamically generated, so those pages cannot be ignored by search engines. Our aim is to generate guidelines to crawl these new, practically infinite, Web sites.

References

1. Raghavan, S., Garcia-Molina, H.: Crawling the hidden web. In: Proceedings of the Twenty-seventh International Conference on Very Large Databases (VLDB), Rome, Italy, Morgan Kaufmann (2001) 129–138
2. Heydon, A., Najork, M.: Mercator: A scalable, extensible web crawler. World Wide Web Conference **2** (1999) 219–229
3. Burke, R.D.: Salticus: guided crawling for personal digital libraries. In: Proceedings of the first ACM/IEEE-CS joint conference on Digital Libraries, Roanoke, Virginia (2001) 88–89
4. Baeza-Yates, R., Castillo, C.: Balancing volume, quality and freshness in web crawling. In: Soft Computing Systems - Design, Management and Applications, Santiago, Chile, IOS Press Amsterdam (2002) 565–572
5. Cho, J., Garcia-Molina, H.: Parallel crawlers. In: Proceedings of the eleventh international conference on World Wide Web, Honolulu, Hawaii, USA, ACM Press (2002) 124–135
6. Chakrabarti, S.: Mining the Web. Morgan Kaufmann Publishers (2003)
7. Diligenti, M., Gori, M., Maggini, M.: A unified probabilistis framework for web page scoring systems. IEEE Transactions on Knowledge and Data Engineering **16** (2004) 4–16

8. Page, L., Brin, S., Motwani, R., Winograd, T.: The pagerank citation algorithm: bringing order to the web. In: Proceedings of the seventh conference on World Wide Web, Brisbane, Australia (1998)

9. Henzinger, M., Heydon, A., Mitzenmacher, M., Najork, M.: On near–uniform url sampling. In: Proceedings of the Ninth Conference on World Wide Web, Amsterdam, Netherlands, Elsevier Science (2000) 295–308

10. Najork, M., Wiener, J.L.: Breadth-first crawling yields high-quality pages. In: Proceedings of the Tenth Conference on World Wide Web, Hong Kong, Elsevier Science (2001) 114–118

11. Cho, J., Garcia-Molina, H.: Synchronizing a database to improve freshness. In: Proceedings of ACM International Conference on Management of Data (SIGMOD), Dallas, Texas, USA (2000) 117–128

12. Henzinger, M.: Hyperlink analysis for the web. IEEE Internet Computing **5** (2001) 45–50

13. Haigh, S., Megarity, J.: Measuring web site usage: Log file analysis. Network Notes (1998)

14. Tauscher, L., Greenberg, S.: Revisitation patterns in world wide web navigation. In: Proceedings of the Conference on Human Factors in Computing Systems CHI'97. (1997)

15. Tanasa, D., Trousse, B.: Advanced data preprocessing for intersites Web usage mining. IEEE Intelligent Systems **19** (2004) 59–65

16. Tan, P.N., Kumar, V.: Discovery of web robots session based on their navigational patterns. Data Mining and Knowledge discovery **6** (2002) 9–35

17. Huberman, B.A., Pirolli, P.L.T., Pitkow, J.E., Lukose, R.M.: Strong regularities in world wide web surfing. Science **280** (1998) 95–97

18. Adar, E., Huberman, B.A.: The economics of web surfing. In: Poster Proceedings of the Ninth Conference on World Wide Web, Amsterdam, Netherlands (2000)

19. Levene, M., Borges, J., Loizou, G.: Zipf's law for web surfers. Knowledge and Information Systems **3** (2001) 120–129

20. Lukose, R.M., Huberman, B.A.: Surfing as a real option. In: Proceedings of the first international conference on Information and computation economies, ACM Press (1998) 45–51

21. Liu, J., Zhang, S., Yang, J.: Characterizing web usage regularities with information foraging agents. IEEE Transactions on Knowledge and Data Engineering **16** (2004) 566 – 584

22. Cooley, R., Mobasher, B., Srivastava, J.: Data preparation for mining world wide web browsing patterns. Knowledge and Information Systems **1** (1999) 5–32

23. Catledge, L., Pitkow, J.: Characterizing browsing behaviors on the world wide web. Computer Networks and ISDN Systems **6** (1995)

Do Your Worst to Make the Best: Paradoxical Effects in PageRank Incremental Computations*

Paolo Boldi[1], Massimo Santini[2], and Sebastiano Vigna[1]

[1] Dipartimento di Scienze dell'Informazione, Università degli Studi di Milano,
via Comelico 39/41, I-20135 Milano, Italy
[2] Università di Modena e Reggio Emilia,
via Giglioli Valle I-42100 Reggio Emilia, Italy

Abstract. Deciding which kind of visit accumulates high-quality pages more quickly is one of the most often debated issue in the design of web crawlers. It is known that breadth-first visits work well, as they tend to discover pages with high PageRank early on in the crawl. Indeed, this visit order is much better than depth first, which is in turn even worse than a random visit; nevertheless, breadth-first can be superseded using an omniscient visit that chooses, at every step, the node of highest PageRank in the frontier.

This paper discusses a related, and previously overlooked, measure of effectivity for crawl strategies: whether the graph obtained after a partial visit is in some sense representative of the underlying web graph as far as the computation of PageRank is concerned. More precisely, we are interested in determining how rapidly the computation of PageRank over the visited subgraph yields relative ranks that agree with the ones the nodes have in the complete graph; ranks are compared using Kendall's τ.

We describe a number of large-scale experiments that show the following paradoxical effect: visits that gather PageRank more quickly (e.g., highest-quality-first) are also those that tend to miscalculate PageRank. Finally, we perform the same kind of experimental analysis on some synthetic random graphs, generated using well-known web-graph models: the results are almost opposite to those obtained on real web graphs.

1 Introduction

The search for the better crawling strategy (i.e., a strategy that gathers early pages of high quality) is by now an almost old research topic (see, e.g., [1]). Being able to collect quickly high-quality pages is one of the major design goals of a crawler; this issue is particularly important, because, as noted in [2], *even*

---

* This work has been partially supported by MIUR COFIN "Linguaggi formali e automi: metodi, modelli e applicazioni" and by a "Finanziamento per grandi e mega attrezzature scientifiche" of the Università degli Studi di Milano.

S. Leonardi (Ed.): WAW 2004, LNCS 3243, pp. 168–180, 2004.

after crawling well over a billion pages, the number of uncrawled pages still far exceeds the number of crawled pages.[1]

Usually, a basic measure of quality is PageRank [3], in one of its many variants. Hence, as a first step we can compare two strategies by looking at how fast the cumulative PageRank (i.e., the sum of the PageRanks of all the visited pages up to a certain point) grows over time. Of course, this comparison can only be performed after the end of the crawl, because one needs the whole graph to compute PageRank.

We list a number of classical visit strategies:

- *Depth-first* order: the crawler chooses the next page as the *last* that was added to the frontier; in other words, the visit proceeds in a LIFO fashion.
- *Random* order: the crawler chooses randomly the next page from the frontier.
- *Breadth-first* order: the crawler chooses the next page as the *first* that was added the frontier; in other words, the visit proceeds in a FIFO fashion.
- *Omniscient* order (or quality-first order): the crawler uses a queue prioritised by PageRank [1]; in other words, it chooses to visit the page with highest quality among the ones in the frontier. This visit is meaningless unless a previous PageRank computation of the entire graph has been performed *before* the visit, but it is useful for comparisons. A variant of this strategy may also be adopted if we have already performed a crawl and so we have the (old) PageRank values of (at least some of the) pages.

Both common sense and experiments (see, in particular, [4]) suggest that the above visits accumulate PageRank in a growingly quicker way. This is to be expected, as the omniscient visit will point immediately to pages of high quality. The fact that breadth-first visit yields high-quality pages was noted in [5].

There is, however, another and also quite relevant problem, which has been previously overlooked in the literature: if we assume that the crawler has no previous knowledge of the web region it has to crawl, it is natural that it will try to detect page quality during the crawl itself, by computing PageRank on the region it has just seen. We would like to know whether doing so it will obtain reasonable results or not. This question is even more urgent than it might appear at first, because we will most probably stop the crawl at a certain point anyway, and use the graph we have crawled to compute PageRank.

To answer this question, of course, we must establish a measure of how good is a PageRank approximation computed on a subgraph. Comparing directly PageRank values would be of course meaningless because of normalisation. However, we can compare two orders:

- the order induced on the subgraph by PageRank (as computed on the subgraph itself);

---

[1] In the present paper by "uncrawled page" we are referring only to those pages that actually exist and will be crawled sometimes in the future, did the crawl go on forever.

– the order induced on the whole graph by PageRank; of course, this order
must be restricted to the nodes of the subgraph.

In this paper, we use Kendall's τ as a measure of concordance between the
two ranked lists (a similar approach was followed in [6] to motivate the usefulness of a hierarchical algorithm to compute PageRank). The results we obtain
are paradoxical: *many strategies that accumulate PageRank quickly explore subgraphs with badly correlated ranks*, and viceversa. Even more interestingly, these
behaviours have not been reproduced by random graphs generated with two
well-known models.

2 Kendall's τ and Its Computation

There are many correlation indices that can be used to compare orders[2]. One of
the most widely used and intuitive is Kendall's τ; this classical nonparametric
correlation index has recently received much attention within the web community
for its possible applications to rank aggregation [8, 9, 10] and for determining the
convergence speed in the computation of PageRank [11]. Kendall's τ is usually
defined as follows[3]:

Definition 1 ([12], pages 34–36). *Let $r_i, s_i \in \mathbf{R}$ ($i = 1, 2, \ldots, n$) be two
rankings. Given a pair of distinct indices $1 \leq i, j \leq n$, we say that the pair is:*

- *concordant iff $r_i - r_j$ and $s_i - s_j$ are both nonzero and have the same sign;*
- *discordant iff $r_i - r_j$ and $s_i - s_j$ are both nonzero and have opposite signs;*
- *an r-tie iff $r_i - r_j = 0$;*
- *an s-tie iff $s_i - s_j = 0$;*
- *a joint tie iff $r_i - r_j = s_i - s_j = 0$.*

Let C, D, T_r, T_s, J be the number of concordant pairs, discordant pairs, r-ties,
s-ties and joint ties, respectively; let also $N = n(n-1)/2$. Of course $C + D +
T_r + T_s - J = N$. Kendall's τ of the two rankings is now defined by

$$\tau = \frac{C - D}{\sqrt{(N - T_r)(N - T_s)}}.$$

Kendall's τ is always in the range $[-1, 1]$: $\tau = 1$ happens iff there are no non-
joint ties and the two total orders induced by the ranks are the same; $\tau = -1$,
conversely, happens iff there are no non-joint ties and the two total orders are
opposite of each other; thus, $\tau = 0$ can be interpreted as lack of correlation.

Knight's Algorithm. Apparently, evaluating τ on large data samples is not so
common, and there is a remarkable scarcity of literature on the subject of com-

---

[2] For a thorough account on this topic, consult for example [7].
[3] There are actually various subtly different definitions of τ, depending on how ties
should be treated. The definition we use here is usually referred to as τ_b.

puting τ in an efficient way. Clearly, the brute-force $O(n^2)$ approach is easy to implement, but inefficient. Knight [13] presented in the sixties an $O(n \log n)$ algorithm for the computation of τ, but the only implementation we are aware of belongs to the SAS system.

Because of the large size of our data, we decided to write a direct, efficient implementation of Knight's algorithm, with a special variant needed to treat ties as required by our definition. This variant was indeed briefly sketched at the end of Knight's original paper, but details were completely omitted[4].

First of all, we sort the indices $\{1, 2, \ldots, n\}$ using r_i as first key, and s_i as secondary key. More precisely, we compute a permutation π of the indices such that $r_{\pi(i)} \leq_{\pi(j)}$ whenever $i \leq j$ and, moreover, if $r_{\pi(i)} = r_{\pi(j)}$ but $s_{\pi(i)} < s_{\pi(j)}$ then $i < j$.

Once the elements have been sorted, a linear scan is sufficient to compute T_r: a maximal sequence of indices i, $i+1$, ..., j such that $r_{\pi(i)} = r_{\pi i+1} = \cdots = r_{\pi(j)}$ determines exactly $\binom{j-i+1}{2}$ r-ties. Moreover, with another linear scan, and in a completely analogous way, one can compute J (this time, one looks for maximal intervals where both r and s are constant). After the computation of D (described below), the indices will be sorted using s_i, so we shall able to compute in a similar way T_s.

The knowledge of D is now sufficient to deduce τ; the computation of D is indeed the most interesting part of Knight's algorithm. Remember that all indices are at this point sorted using r_i as first key, and s_i as secondary key. We apply a stable merge sort to all indices using s_i as key. Every time, during a merge, we move an item *forward*, we increase the number of discordant pairs by the number of skipped items. Thus, for instance, passing from the order 2, 1, 0, 3 to natural order requires first ordering two sublists of length 2, getting to 1, 2, 0, 3 and increasing by 1 the number of discordant pairs, and then moving 0 to the first position, getting to 0, 1, 2, 3 and bringing the overall number to 3. Note that the same amount can be more easily calculated using a bubble sort: you just need two exchanges to move 0 to the front, and then one to move 1 to its final position (the idea of using bubble sort appears in Kendall's original paper [14]), but you cannot expect to run bubble sort on a graph with 100 million nodes.

3   Measuring the Quality of a Page with PageRank

PageRank [15] is one of the best-known methods to measure page quality; it is a static method (in that it does not depend on a specific query but rather it measures the absolute authoritativeness of each page), and it is based purely on the structure of the links, or, if you prefer, on the web graph. As it was recently noticed [2], PageRank is actually a set of ranking methods that depends on some parameters and variants; its most common version can be described as the

---

[4] The complete Java code of our implementation of Knight's algorithm above is available under the GNU General Public License.

behaviour of a random surfer walking through the web, and depends on a single parameter $\alpha \in (0, 1)$ (the *dumping factor*).

The random surfer, at each step t, is in some node p_t of the graph. The node p_{t+1} where the surfer will be in the next step is chosen as follows:

- if p_t had no outgoing arcs, p_{t+1} is chosen uniformly at random among all nodes;
- otherwise, with probability α, one of the arcs going out of p_t is chosen (with uniform distribution), and p_{t+1} is the node where the arc ends; with probability $1 - \alpha$, p_{t+1} is once more chosen uniformly among all nodes.

The PageRank of a given node x is simply the fraction of time that the random surfer spent in x, and will be denoted by $\mathrm{PR}_G(x)$.

Another, maybe more perspicuous, way of defining PageRank is the following. Let $A = (a_{ij})$ be an $n \times n$ matrix (n being the number of nodes in the graph), defined as follows:

- if i has no outgoing arcs, $a_{ij} = 1/n$;
- if i has $d > 0$ outgoing arcs, and (i, j) is an arc, then $a_{ij} = \alpha/d + (1 - \alpha)/n$;
- if i has $d > 0$ outgoing arcs, but (i, j) is not an arc, then $a_{ij} = (1 - \alpha)/n$.

Now A turns out to be an aperiodic and irreducible stochastic matrix, so there is exactly one probability vector r satisfying

$$Ar = r.$$

The PageRank of a node x, $\mathrm{PR}_G(x)$, is exactly r_x. The computation of PageRank is a rather straightforward task, that can be accomplished with standard tools of linear algebra [16, 11]. The dumping factor α influences both the convergence speed and the results obtained, but it is by now customary to choose $\alpha = 0.85$.

4   Using Kendall's τ to Contrast Crawl Strategies

The rank assigned by PageRank to a page is not important *per se*; it is the relative order of pages with respect to PageRank that is actually interesting for search engines. This observation will guide us in what follows.

Consider the following typical scenario: G, the "real" web graph (unknown to the crawler), has N nodes; we perform some crawl of G, which determines a certain node order (the visit order) $x_1, x_2, \ldots, x_N$. For each $n = 1, 2, \ldots, N$, let $G_n = G[\{x_1, x_2, \ldots, x_n\}]$ be the subgraph of G induced by the first n visited nodes. In other words, G_n is the graph collected at the n-th step of the traversal.

In a real-world crawler, we will most probably stop the crawl at a certain point, say, after n pages have been crawled, typically with $n \ll N$. If you run PageRank on G_n you obtain page rank values that will usually not coincide with the rank values that those pages have in G. Even worse, their *relative order* will be different.

Let τ_n be defined as Kendall's τ computed on $\mathrm{PR}_{G_n}(x_1)$, $\mathrm{PR}_{G_n}(x_2)$, $\ldots$, $\mathrm{PR}_{G_n}(x_n)$ and $\mathrm{PR}_G(x_1)$, $\mathrm{PR}_G(x_2)$, $\ldots$, $\mathrm{PR}_G(x_n)$. Clearly, $\tau_N = 1$, as at the end of the crawl G is entirely known, but the sequence $\tau_1, \tau_2, \ldots, \tau_N$ (hereafter referred to as the τ *sequence of the visit*) only depends on the visit order; it is not necessarily monotonic, but its behaviour can used as a measure of how good (or bad) is the chosen strategy with respect to the way PageRank order is approximated during the crawl. The main goal of this paper is to present experimental results comparing τ sequences produced by different visit strategies. We shall also consider τ sequences of subgraphs not obtained through visits, and use them for comparison.

Observe that a τ sequence has nothing to do with the convergence speed of PageRank, but rather with its tolerance to graph modifications, or, if you prefer, with its stability. There is a rich stream of research (see, for example, [17, 18, 19, 20, 21, 22]) concerning the robustness of PageRank with respect to graph modifications (node and/or link perturbation, deletion and insertion). Many authors observed, in particular, that PageRank is quite stable under graph modifications [20], at least when the changing pages do not have high PageRank; even removing a large portion of pages, as many experiments suggest [18], turns out to have small influence on PageRank. Some papers also propose variants of the PageRank algorithm [19] that can be computed adaptively and robustly with respect to graph evolution.

However, most of these studies (with the exception of [22]) measure the stability of PageRank by computing the distance (usually, under L^1 or L^2 norm) between the PageRank vectors. However interesting this metric might be, it is not directly related to our measure. One might indeed object that τ is not a completely fair way to analyse PageRank computation: after all, two PageRank vectors may turn out to be poorly correlated only because of small value fluctuations in pages with low PageRank. Otherwise said, since τ is not continuous, small perturbations in the PageRank vector may have disastrous effects.

Nevertheless, the real-world usage of PageRank is, by its very nature, discontinuous—PageRank is essentially used only to rank pages, its real numerical value being irrelevant. Moreover, even if fluctuations may only concern low-ranked pages, thus not affecting the mutual order of the top nodes, this fact should not be underestimated; if a query is satisfied only by a set of pages of low rank, their relative order will be for the user as worthy as the order of the top nodes. The latter effect is noticeable not only for some esoteric queries, but typically also for relevant queries of interest only to a small community: you will not expect that a query like *group torsion* or *analog editing* is satisfied by pages with high PageRank, but the relative order of the pages that satisfy those queries is certainly meaningful.

5   Experimental Setup

Graphs. We based our experiments on four web graphs:

- a 41,291,594-nodes snapshot of the Italian domain .it, performed with UbiCrawler [4] (later on referred to as the *Italian graph*);

- a 118,142,155-nodes graph obtained from the 2001 crawl performed by the WebBase crawler [23] (later on referred to as the *WebBase graph*);
- a 10,000,000-nodes synthetic graph generated using the Copying Model proposed in [6], with suitable parameters and rewiring (later on referred to as the *copying-model graph*);
- a 10,000,000-nodes synthetic graph generated using the Evolving Network Model proposed in [24], with suitable parameters and rewiring (later on referred to as the *evolving-model graph*).

The two synthetic graphs are much smaller than the others, because by their random nature there would be no gain in using a larger node set.

For each graph, we computed the strongly-connected components, and discarded the nodes that were not reachable from the giant component: we decided to do so because otherwise a single visit would not collect the whole graph, and the visit order would not depend only on the visit strategy and on a single starting node. The resulting graph sizes were 41,291,594 for the Italian graph, 95,688,816 for the WebBase graph, 4,863,948 for the copying-model graph and 5,159,894 for the evolving-model graph.

Visit Seeds. For each of the four graphs produced as above, we performed visits starting from three different seeds (a.k.a. starting URLs). We decided to perform visits from the highest-ranked, lowest-ranked and median-ranked node within the giant component (choosing seeds within the giant component guarantees that the whole graph will be collected at the end of the visit).

Visit Strategies. For each of the graphs and every visit seed, we performed four different kinds of visits: BF (breadth-first traversal), DF (depth-first traversal), RF (random-first traversal: the next node to be visited is chosen at random from the frontier), QF (quality-first traversal: the next node to be visited is the one having highest PageRank in the frontier; here, by PageRank we mean PageRank in the real graph, not in the subgraph), and IQF (inverse-quality-first traversal: in this case, we choose instead the node having the *lowest* PageRank in the frontier).

Sampling. Computing PageRank and Kendall's τ on a large graph is a heavy computation task. Due to the large size of our data, and to the large number of different visits, we decided to spread logarithmically our sample points, so to avoid sampling too many times very large graphs. In general, samples are multiplicatively spaced at least by $\sqrt{2}$, albeit in several cases we used a thicker sampling. Note also that when the subgraph sizes exceed half of the original graph size, all curves become smooth, so using denser samples in that region is probably useless.

Computational Time. We performed our PageRank and τ computation on five dual-processor PCs and two multi-processor servers (the latter were essential in handling the largest samples). Overall, we used about 1600 hours of CPU user time.

Tools. All our tests were performed using the WebGraph framework [25, 26]. Moreover, at the address http://webgraph-data.dsi.unimi.it/, are available both the Italian and the WebBase graphs (recall, however, that we only used the portion of WebBase that can be reached from the giant component). The two synthetic graphs were produced using COSIN [27].

6   Experimental Results

Comparing τ Sequences. Figure 1 represents the τ sequences obtained during the visits of the four graphs, starting from the node with highest PageRank within the giant component (in the Appendix, you can find similar graphs for other seeds). In each graph, we also plotted the τ sequence RAND corresponding to a random node order (note that this is quite different from a random-first traversal, because it is not a graph visit); additionally, the diagrams contain an

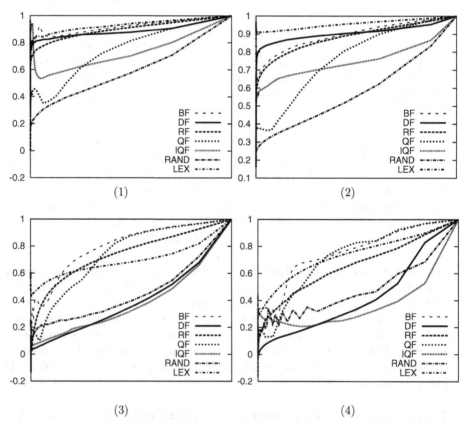

Fig. 1. τ sequences for (1) the Italian graph, (2) the WebBase graph, (3) the copying-model graph and (4) the evolving-model graph, starting from the highest-ranked node in the giant component

extra line, called LEX, whose meaning will be explained in the next section. The horizontal axis represents time (or, more precisely, the number of nodes that have been visited), whereas the vertical axis represents the value of τ.

Cumulative PageRank. For comparison, we computed (Figure 2) the cumulative PageRank (the sum of all PageRanks of the pages collected so far) during the visits of the Italian graph.

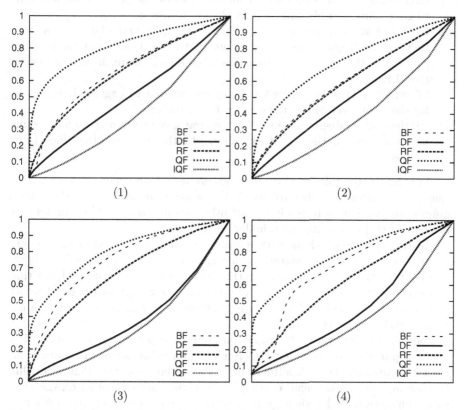

Fig. 2. Cumulative PageRank obtained during a visit of (1) the Italian graph, (2) the WebBase graph, (3) the copying-model graph and (4) the evolving-model graph, starting from the highest-ranked node in the giant component

7   Interpretation of the Experimental Results

Cumulative PageRank. The graph of cumulative PageRank (Figure 2) confirms other experiments reported by the literature: if you want to collect pages with high PageRank, you should use a breadth-first (BF) traversal; this strategy is overcome only by the omniscient quality-first QF. Obviously, using an inverse-quality-first traversal (IQF) produces the poorest results. Also, depth-first DF is only marginally better than IQF, whereas, a bit surprisingly, RF is very good, even though worst than BF. Note that these results, most of which are well-known or folklore, do not appear to be much influenced by the seed.

PageRank: Locality and Collective Nature. We now turn to the study of the τ sequences (Figure 1). Looking at the diagrams of the Italian and of the WebBase graphs, we observe the following facts.

1. Even though τ sequences do not increase monotonically, they anyway tend to grow quite steadily, after a brief chaotic transient.
2. DF and BF are, in most cases, comparable; often, DF is even better. Moreover, they are both quite close to RF. Note that these behaviours are in sharp contrast with the case of cumulative PageRank, where BF was far better than DF.
3. QF, which used to be the best strategy for cumulative PageRank, is now by far the worst possible strategy. Even IQF is better!
4. All the visit strategies considered are anyway more τ-effective than collecting nodes at random: the line of RAND is constantly below all other lines, in all graphs.

Apparently, the rôles of visit strategies are now completely scrambled. Note that the situation is rather robust: it is the same for all seeds and for both graphs, and we indeed tested it against other, smaller real-world examples (not presented here) obtaining quite similar behaviours.

To try to understand the very reasons behind the observed facts, it might be helpful to look at the diagrams in logarithmic (horizontal) scale (not shown here). Under this scale, one observes that all visits present an initial burst that is absent in RAND, whose growth is quite steady. Our explanation for this fact is that, at the beginning of a visit, the crawl tends to remain confined within a single website (or a set of websites).

Indeed, a recent work [28] makes an analysis quite similar to ours with a different purpose: to prove that PageRank computation can be made more efficient by computing local PageRanks first. The authors motivate their algorithm by showing that local PageRanks agree when they are computed on the local subgraph instead of the whole graph, and they use (a version of) Kendall's τ to measure agreement (they perform their experiments on a small 683,500 pages snapshot Stanford/Berkeley site).

Essentially, inside a single site PageRank depends heavily on the local links. This phenomenon, that we might call *PageRank locality*, explains why any visit performs better than a random selection of nodes.

With the aim of obtaining an empirical confirmation for our conjecture of PageRank locality, we tried another experiment. We computed the τ sequence of subsets of nodes collected in lexicographic order; in other words, we started with the first URL (in lexicographic order), and proceeded collecting URLs in this way. The resulting τ sequences are shown in the diagram as LEX: since LEX is a most local way of collecting nodes, we expect the τ sequence to be distinctly good. And, indeed, LEX lies always above all other lines.

Locality can also explain the success of DF (which beats BF in the first half of the WebBase graph). Depth-first visits tend to wander a while in a site, and then, as soon as a cross-site link is chosen, jump to another site, wander a while, and so on. This behaviour is closer to LEX than to a breadth-first visit.

However, locality cannot explain other phenomena, such as how an inverse quality visit can beat for a long time an omniscient visit. We believe that the paradoxical behaviour of priority-based visits can only be explained by the *collective nature* of PageRank. To have a good τ-correlation with PageRank, besides being local, a subgraph must be also representative of the collection of pages, that is, *sufficiently random* (with respect to PageRank ordering). Said otherwise, *high authoritativeness cannot be achieved without the help of minor pages.* Selecting blindly the most authoritative pages leads to a world where authority is subverted[5].

Random Models. The graphs for the two chosen random models and are quite disappointing. The PageRank-cumulation behaviour (Figure 2) follows in a relatively close way what happens for real graphs (albeit the models are somehow too good to be true—breadth-first visits beat by far and large random visits, which is not true on real graphs). When, however, we turn to τ-sequences (Figure 1), we find marked differences. For example, DF works much more poorly than it does in real web graphs, whereas QF seems very powerful while it is not. A partial explanation for this is that the chosen models do not take locality into account, hence failing to show the behaviours discussed above. Certainly there is a lot of room for improvement.

8 Conclusions and Further Work

Our results, as the title suggests, are somehow paradoxical: strategies that work quite well when you want to accumulate pages with high quality in a short time tend to behave rather poorly when you try to approximate PageRank on a partial crawl, and viceversa. Our explanation for this paradox is PageRank locality. Of course, the results presented in this paper are still very preliminary, and lack any theoretical investigation. Finally, we address some issues about our methodology and some possible further direction.

- *Treatment of dangling links.* In this paper, we computed the PageRank of a subgraph ignoring the fact that some nodes of the subgraph would contain many outlinks that are not there simply because our crawl is partial. Some authors recently suggested [2] to modify the definition of PageRank to take this phenomenon into account. It would be interesting to know whether their version of PageRank makes things better or not.
- *Comparing only top nodes.* Another possible direction is considering a different metric that does not take into account the whole order, but only the

---

[5] Again, even the opposite choice—selecting the pages with lowest authority—works sometimes better, which should lead us to reconsider gossiping versus good newspapers as a mean for selecting authoritative sources of information.

relative order of the k top elements, where k is either constant, or a fixed ratio of nodes; this approach might also be modified so to use a definition of τ that works also when the orders are not defined over the same set of nodes, like the ones given in [8]. Note, however, that in real-world applications the whole order is very important (even more important than the order of high-quality nodes, as explained at the end of Section 4).

- *More strategies.* Our results would suggest the implementation of crawl strategies that work effectively with respect to both PageRank and τ. A natural candidate strategy would be a mixed order, using a depth-first intra-site visit, and operating breadth-first for extra-site links. This strategy would be quite natural for a parallel or distributed crawler, in that every worker might proceed in depth-first order within a site (respecting, of course, the usual politeness assumptions) whereas out-of-site links might be treated in a breadth-first manner: UbiCrawler [4] was originally designed to work with this strategy. Another interesting strategy could be performing a crawl in lexicographic order (the next node to be visited is the first node of the frontier in lexicographic order).
- *More random models.* The two graph models we have tried presented behaviours that did not quite match with real-world graphs. A natural investigation would be trying to discover if other more sophisticated models work better.

References

1. Cho, J., García-Molina, H., Page, L.: Efficient crawling through URL ordering. Computer Networks and ISDN Systems **30** (1998) 161–172
2. Eiron, N., McCurley, K.S., Tomlin, J.A.: Ranking the web frontier. In: Proceedings of the 13th conference on World Wide Web, ACM Press (2004) 309–318
3. Page, L., Brin, S., Motwani, R., Winograd, T.: The pagerank citation ranking: Bringing order to the web. Technical report, Stanford Digital Library Technologies Project, Stanford University, Stanford, CA, USA (1998)
4. Boldi, P., Codenotti, B., Santini, M., Vigna, S.: Ubicrawler: A scalable fully distributed web crawler. Software: Practice & Experience **34** (2004) 711–726
5. Najork, M., Wiener, J.L.: Breadth-first search crawling yields high-quality pages. In: Proc. of Tenth International World Wide Web Conference, Hong Kong, China (2001)
6. Kumar, R., Raghavan, P., Rajagopalan, S., Sivakumar, D., Tomkins, A., Upfal, E.: Stochastic models for the web graph. In: Proceedings of the 41st Annual Symposium on Foundations of Computer Science, IEEE Computer Society (2000) 57
7. Kendall, M., Gibbons, J.D.: Rank Correlation Methods. Edward Arnold (1990)
8. Fagin, R., Kumar, R., Sivakumar, D.: Comparing top k lists. In: Proceedings of the fourteenth annual ACM-SIAM symposium on Discrete algorithms, Society for Industrial and Applied Mathematics (2003) 28–36
9. Fagin, R., Kumar, R., McCurley, K.S., Novak, J., Sivakumar, D., Tomlin, J.A., Williamson, D.P.: Searching the workplace web. In: Proceedings of the twelfth international conference on World Wide Web, ACM Press (2003) 366–375

10. Dwork, C., Kumar, R., Naor, M., Sivakumar, D.: Rank aggregation methods for the web. In: Proceedings of the tenth international conference on World Wide Web, ACM Press (2001) 613–622

11. Kamvar, S.D., Haveliwala, T.H., Manning, C.D., Golub, G.H.: Extrapolation methods for accelerating pagerank computations. In: Proceedings of the twelfth international conference on World Wide Web, ACM Press (2003) 261–270

12. Kendall, M.G.: Rank Correlation Methods. Hafner Publishing Co., New York (1955)

13. Knight, W.R.: A computer method for calculating kendall's tau with ungrouped data. Journal of the American Statistical Association **61** (1966) 436–439

14. Kendall, M.G.: A new measure of rank correlation. Biometrika **30** (1938) 81–93

15. Brin, S., Page, L.: The anatomy of a large-scale hypertextual web search engine. Computer Networks **30** (1998) 107–117

16. Haveliwala, T.: Efficient computation of pagerank. Technical report, Stanford University (1999)

17. Lee, H.C., Borodin, A.: Perturbation of the hyper-linked environment. In: Computing and Combinatorics, 9th Annual International Conference, COCOON 2003, Big Sky, MT, USA, July 25-28, 2003, Proceedings. Volume 2697 of Lecture Notes in Computer Science., Springer (2003) 272–283

18. Ng, A.Y., Zheng, A.X., Jordan, M.I.: Stable algorithms for link analysis. In: Proceedings of the 24th annual international ACM SIGIR conference on Research and development in information retrieval, ACM Press (2001) 258–266

19. Abiteboul, S., Preda, M., Cobena, G.: Adaptive on-line page importance computation. In: Proceedings of the twelfth international conference on World Wide Web, ACM Press (2003) 280–290

20. Bianchini, M., Gori, M., Scarselli, F.: Inside pageRank. ACM Transactions on Internet Technologies (2004) To appear.

21. Langville, A.N., Meyer, C.D.: Deeper inside pageRank. Internet Mathematics (2004) To appear.

22. Lempel, R., Moran, S.: Rank stability and rank similarity of link-based web ranking algorithms in authority connected graphs (2004) Information Retrieval (in print); special issue on Advances in Mathematics and Formal Methods in Information Retrieval.

23. Hirai, J., Raghavan, S., Garcia-Molina, H., Paepcke, A.: Webbase: A repository of web pages. In: Proc. of WWW9, Amsterdam, The Netherlands (2000)

24. Albert, R., Barábasi, A.L., Jeong, H.: Diameter of the World Wide Web. Nature **401** (1999)

25. Boldi, P., Vigna, S.: The WebGraph framework I: Compression techniques. In: Proc. of the Thirteenth International World Wide Web Conference, Manhattan, USA (2004) 595–601

26. Boldi, P., Vigna, S.: The WebGraph framework II: Codes for the World–Wide Web. Technical Report 294-03, Università di Milano, Dipartimento di Scienze dell'Informazione (2003) To appear as a poster in *Proc. of DCC 2004*, IEEE Press.

27. Donato, D., Laura, L., Leonardi, S., Milozzi, S.: A library of software tools for performing measures on large networks (2004) http://www.dis.uniroma1.it/~cosin/html_pages/COSIN-Tools.htm.

28. Kamvar, S.D., Haveliwala, T.H., Manning, C.D., Golub, G.H.: Exploiting the block structure of the web for computing pagerank. Technical report, Stanford University (2003)

Communities Detection in Large Networks

Andrea Capocci[1], Vito D.P. Servedio[1,2],
Guido Caldarelli[1,2], and Francesca Colaiori[2]

[1] Centro Studi e Ricerche e Museo della Fisica "E. Fermi",
Compendio Viminale, Roma, Italy
[2] INFM UdR Roma1-Dipartimento di Fisica Università "La Sapienza",
P.le A. Moro 5, 00185 Roma, Italy

Abstract. We develop an algorithm to detect community structure in
complex networks. The algorithm is based on spectral methods and takes
into account weights and links orientations. Since the method detects
efficiently clustered nodes in large networks even when these are not
sharply partitioned, it turns to be specially suitable to the analysis of
social and information networks. We test the algorithm on a large-scale
data-set from a psychological experiment of word association. In this
case, it proves to be successful both in clustering words, and in uncovering
mental association patterns.

1 Communities in Networks

Complex networks are composed by a large number of elements (nodes), organized into sub-communities summing up to form the whole system. The partition
of a complex network has no unique solution, since nodes can be aggregated according to different criteria. However, when dealing with networks an additional
constraint bounds the possible choices: the entire information is coded into the
adjacency matrix, whose elements describe the connection between each pair
of nodes. Detecting communities, therefore, means uncovering the reduced
number of interactions involving many unit elements, given a large set of pair
interactions.

The partition of a graph into communities has a broad range of technological applications (from the detection of polygenic interaction in genetic networks
to the development of effective tools for information mining in communication
networks [1–3]). A perhaps more intriguing scientific interest in network partitioning comes from social sciences, where methods to detect groups within social
networks are employed on a daily basis [4]. Moreover, when used in the analysis
of large collaboration networks, such as company or universities, communities
reveal the informal organization and the nature of information flows through
the whole system [5, 6]. Social networks, however, are usually described by undirected graphs, where links are reciprocal. This somewhat simplifies the task, and
subtler methods are required when considering directed graphs, as shown below.

Despite the potential application of the results, measurements about structures involving more than two nodes in networks mainly concern regular pat-

S. Leonardi (Ed.): WAW 2004, LNCS 3243, pp. 181–187, 2004.
© Springer-Verlag Berlin Heidelberg 2004

terns [7–13], and particularly the clustering coefficient, which counts the number of triangles in a graph.

2 Network Partitioning Algorithm

Indeed, few scientist have developed methods and algorithms able to identify irregularly shaped communities. Traditional methods are divisive, i.e. they are based on the fragmentation of the original network into disconnected components by the removal of links.

2.1 Edge-Betweenness Methods

Recent algorithms [3, 14] are mainly based on the edge betweenness or local analogues of it. Edge betweenness measures how central an edge is: to assess this, one finds the shortest paths between each pair of nodes, and counts the fraction of them which runs over the considered edge. Removing a sufficient number of such edges from the network fragments it into separate parts. The process is iterated until the network is split into cluster including only individual nodes. The NG–algorithm builds a tree or, more exactly, a dendrogram: at each splitting event, nodes are subdivided in families, sub-families and so on.

This methods is widely assumed to give reasonable partition of networks, since the first splittings produce the most basic subgroups, whereas successive splittings gives a classification of the nodes into families at a higher resolution. Based on a similar principle, the method introduced in ref. [14] has the advantage of being faster. Despite its the outcome is independent on how sharp the partitioning of the graph is.

2.2 Spectral Methods

Spectral methods study the adjacency matrix A [15–17], whose generic element a_{ij} is equal to 1 if i points to j and 0 otherwise.

In particular, such methods analyze matrix related to A, such as the Laplacian matrix $K - A$ and the Normal matrix $K^{-1}A$, where K is the diagonal matrix with elements $k_{ii} = \sum_{j=1}^{S} a_{ij}$ and S is the number of nodes in the network. In most approaches, referring to undirected networks, A is assumed to be symmetric.

The Laplacian and Normal spectrum have peculiar shapes. The largest eigenvalue of the Normal spectrum is equal to 1 by definition, since it corresponds to the normalization to one imposed to elements on a same matrix row. But nonprincipal eigenvalues close to 1 gives an insight about the network partitioning, since they correspond roughly to the number of clear components of the graph. A perturbative approach helps to understand this: a block matrix would have an eigenvalue equal to 1 for each matrix block, since the normalization applies to each block as well. If some link across the block is present, this slightly perturbs the spectrum, but eigenvalues close to one are still meaningful if a clear partition is present. A similar argument applies for the Laplacian matrix, but in this case eigenvalues close to 0 are to be looked at.

The eigenvectors associated to the largest eigenvalues of the Laplacian and Normal spetrum have a characteristic structure too: the components corresponding to nodes within the same cluster have very similar values x_i, so that, as long as the partition is sufficiently sharp, the profile of each eigenvector, sorted by components, is step–like. The number of steps in the profile corresponds again to the number of communities.

This is explained by mapping the matrix eigenproblem into a constrained minimization process. With the most general applications in mind, we replace the adjacency matrix A by the weight matrix W, whose elements w_{ij} are assigned the intensity of the link (i, j). We consider undirected graphs first, and then we pass to the most general directed case.

The spectrum of such matrices are related to the minimization of

$$z(\mathbf{x}) = \frac{1}{2} \sum_{i,j=1}^{S} (x_i - x_j)^2 w_{ij} \,, \tag{1}$$

where x_i are values assigned to the nodes, with some constraint on the vector $\mathbf{x}$, expressed by

$$\sum_{i,j=1}^{S} x_i x_j m_{ij} = 1 \,, \tag{2}$$

where m_{ij} are elements of a given symmetric matrix M.

Writing $\frac{dz}{dx_i} = 0$ for all x_i in vector formalism, given the constraint (2) leads to equation

$$(D - W)\mathbf{x} = \mu M \mathbf{x} \,, \tag{3}$$

where D is the diagonal matrix $d_{ij} = \delta_{ij} \sum_{k=1}^{S} w_{ik}$, and μ is a Lagrange multiplier.

The constraint Matrix M determine the kind of eigenproblem to solve. Choosing $M = D$ to $D^{-1}W\mathbf{x} = \mu\mathbf{x}$ (related to the generalized Normal spectrum), while $M = 1$ leads to $(D - W)\mathbf{x} = \mu\mathbf{x}$ (corresponding to the Laplacian case).

The absolute minimum corresponds to the trivial eigenvector, which is constant. The other stationary points correspond to eigenvectors where components associated to well connected nodes assume similar values.

As an example, we show in Fig. 2 the step-like profile of the second eigenvectors of $D^{-1}W$ for the simple graph shown in Fig. 1 with $S = 19$ nodes, where random weights between 1 and 10 were assigned to the links.

Yet, studying the eigenvector profile is meaningful only if such a sharp partition exists. In most cases, especially in large networks systems, eigenvector profiles are far too complicate to detect steps, and components do not cluster around a few values. Nevertheless, the minimization problem still applies, so that clustered node correspond to components with similar values in many eigenvectors.

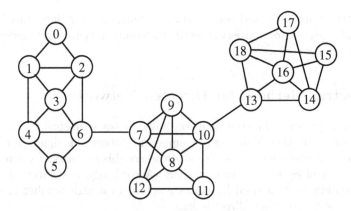

Fig. 1. Network employed as an example, with $S = 19$ and random weights between 1 and 10 assigned to the links. Three clear clusters appear, composed by nodes $0 - 6$, $7 - 12$ and $13 - 19$

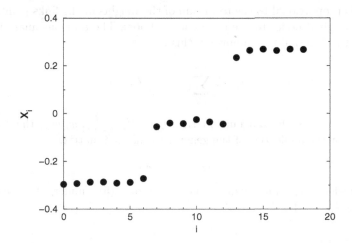

Fig. 2. Values of the 2nd eigenvector components for matrix $D^{-1}W$ relative to the graph depicted in figure 1

By measuring the correlation

$$r_{ij} = \frac{\langle x_i x_j \rangle - \langle x_i \rangle \langle x_j \rangle}{[(\langle x_i^2 \rangle - \langle x_i \rangle^2)(\langle x_j^2 \rangle - \langle x_j \rangle^2)]^{\frac{1}{2}}} \,, \tag{4}$$

where the average $\langle \cdot \rangle$ is over the first few nontrivial eigenvectors, one is then able to detect such a rigid displacement of the components across different eigenvectors.

The quantity r_{ij} measures the community closeness between node i and j. Though the performance may be improved by averaging over more and more eigenvectors, with increased computational effort, we find that indeed a small number of eigenvectors suffices to identify the community to which nodes belong, even in large networks.

3   Spectral Methods for Directed Networks

When dealing with a directed network, links do not correspond to any equivalence relation. Rather, pointing to common neighbors is a significant relation, as suggested in the sociologists' literature where this quantity measures the so-called *structural equivalence* of nodes [18]. Accordingly, in a directed network, clusters should be composed by nodes pointing to a high number of common neighbors, no matter their direct linkage.

To detect the community structure in a directed network, we therefore replace, in the previous analysis, the matrix W by matrix $Y = WW^T$. This corresponds to replacing the directed network with an undirected weighted network, where nodes pointing to common neighbors are connected by a link whose intensity is proportional to the total sum of the weights of the links pointing from the two original nodes to the common neighbors. The previous analysis wholly applies: the function to minimize in this case is

$$y(\mathbf{x}) = \sum_{ijl}^{1,S} (x_i - x_j)^2 w_{il} w_{jl} . \tag{5}$$

Defining Q as the diagonal matrix $q_{ij} = \delta_{ij} \sum_{lj=1}^{S} w_{il} w_{jl}$, the eigenvalue problem for the analogous of the generalized normal matrix,

$$Q^{-1} Y \mathbf{x} = \lambda \mathbf{x} \tag{6}$$

is equivalent to minimizing the function (5) under the constraint $\sum_{ijl=1}^{S} x_i x_j q_{ij} = 1$.

4   A Test: The Word Association Network

To test this spectral correlation-based community detection method on a real complex network, we apply the algorithm to data from a psychological experiment reported in reference [19]. Volunteering participants to the research had to respond quickly by freely associating a word (response) to another word given as input (stimulus), extracted by a fixed subset. Scientists conducting the research have recorded all the stimuli and the associated responses, along with the occurrence of each association. As in ref. [20], we build a network were words are nodes, and directed links are drawn from each stimulus to the corresponding responses, assuming that a link is oriented from the stimulus to the response. The resulting network includes $S = 10616$ nodes, with an average in-degree equal to about 7. Taking into account the frequency of responses to a given stimulus, we construct the weighted adjacency matrix W. In this case, passing to the matrix Y means that we expect stimuli giving rise to the same response to be correlated.

The word association network is an ideal test case for our algorithm, since words are naturally associated by their meaning, so that the performance of our method emerges immediately emerges at glance, when one looks at words falling in the same cluster.

However, in such large databases a partition is not defined, there Rather, one is interested in finding groups of highly correlated nodes, or groups of nodes highly connected to a given one. Table 1 shows the most correlated words to three test-words. The correlation are computed by averaging over just 10 eigenvectors of the matrix $Q^{-1}Y$: the results appear to be quite satisfactory, already with this small number of eigenvectors.

Table 1. The words most correlated to *science*, *literature* and *piano* in the eigenvectors of $Q^{-1}WW^T$. Values indicate the correlation

science	1	literature	1	piano	1
scientific	0.994	dictionary	0.994	cello	0.993
chemistry	0.990	editorial	0.990	fiddle	0.992
physics	0.988	synopsis	0.988	viola	0.990
concentrate	0.973	words	0.987	banjo	0.988
thinking	0.973	grammar	0.986	saxophone	0.985
test	0.973	adjective	0.983	director	0.984
lab	0.969	chapter	0.982	violin	0.983
brain	0.965	prose	0.979	clarinet	0.983
equation	0.963	topic	0.976	oboe	0.983
examine	0.962	English	0.975	theater	0.982

As sown in table 1, the results are quite satisfying: most correlated words have closely related meanings or are directly associated to the test word by a simple relation (synonymy or antinomy, syntactic role, and even by analogous sensory perception).

5   Conclusions

We have introduced a method to detect communities of highly connected nodes within a network. The method is based on spectral analysis and applies to weighted networks. When tested on a real network instance (the records of a psychological experiments) the algorithm proves to be successful: it clusters nodes or, in such case, words, according to natural criteria, and provides an automatic way to extract the most connected sets of nodes to a given one in a set of over 10^4.

The authors thank Miguel-Angel Muñoz and Ramon Ferrer Y Cancho for useful discussion. They acknowledge partial support from the FET Open Project IST-2001-33555 COSIN.

References

1. I. Simonsen, K. A. Eriksen, S. Maslov, K. Sneppen, cond-mat/0312476 (2003), to appear in *Physica A*
2. S.R. Kumar, P. Raghavan, S. Rajagopalan and A. Tomkins, *The VLDB Journal*, 639 (1999).
3. M. Girvan and M. E. J. Newman, *Proc. Natl. Acad. Sci. USA* **99**, 8271 (2002).
4. M. E. J. Newman, *SIAM Review* **45**, 167 (2003).
5. B. Huberman, J. Tyler and D. Wilkinson, in *Communities and technologies*, M. Huysman, E. Wegner and V. Wulf, eds. Kluwer Academic (2003).
6. R. Guimerà, L. Danon, A. Diaz-Guilera, F. Giralt and A. Arenas, *Phys. Rev. E* **68** 065103 (2003)
7. R. Albert and A.-L. Barabási, *Rev. Mod. Phys.* **74**, 47 (2002).
8. S. N. Dorogovtsev and J. F. F. Mendes, *Adv. in Phys.* **51**, 1079 (2002).
9. J.P. Eckmann, E. Moses, *PNAS* **99** (9), 5825 (2002).
10. G. Bianconi and A. Capocci, *Phys. Rev. Lett.* **90** ,078701 (2003).
11. G. Caldarelli, R. Pastor-Satorras and A. Vespignani, cond-mat/0212026 (2002).
12. A. Capocci, G. Caldarelli, P. De Los Rios, *Phys. Rev. E* **68** 047101 (2003).
13. G. Caldarelli, A. Capocci, P. De Los Rios, M.A. Muñoz, *Phys. Rev. Lett.* **89** 258702 (2002).
14. F. Radicchi, C. Castellano, F. Cecconi, V. Loreto and D. Parisi, submitted for publication, preprint cond-mat/0309488
15. K. M. Hall, *Management Science* **17**, 219 (1970).
16. A.J. Seary and W.D. Richards, W.D. *Proceedings of the International Conference on Social Networks Volume 1: Methodology*, 47 (1995).
17. J. Kleinberg, *Journal of the ACM* **46** (5) 604 (1999).
18. M. E. J. Newman, *Eur. Phys. J. B*, in press.
19. M. Steyvers, J. B. Tenenbaum, preprint cond-mat/0110012, submitted for publication.
20. L. Da Fontoura Costa, preprint cond-mat/0309266, submitted for publication

Author Index